ENCYCLOPAEDIA OF
ENVIRONMENTAL SCIENCE

Encyclopaedia of

Environmental Science

Sanjoy Chatterjee

ANMOL PUBLICATIONS PVT. LTD.
NEW DELHI - 110 002 (INDIA)

ANMOL PUBLICATIONS PVT. LTD.
H.O.: 4374/4B, Ansari Road, Daryaganj
New Delhi - 110 002
Ph.: 23278000, 23261597
B.O.: No. 1015, Ist Main Road, BSK III Stage
III Phase, III Block
Bangalore - 560 085 (India)
Visit us at: www.anmolpublications.com

Encyclopaedia of Environmental Science

First Published, 2006

ISBN 81-261-3037-7

PRINTED IN INDIA

Printed at Mehra Offset Press, Delhi

Preface

The study of Environmental science influences almost all sphere of life. Today the society is well aware of its environmental problems. People from numerous disciplines deal with environmental problems everyday, so studies in this field have become a conventional phenomenon. This volume contains terms covering vast range of subjects like acid rain, CFCs, greenhouse effect, environmental pollution, climatology, meteorology, geomorphology, pollution, climate, forestation, desertification, endangered species, environmental impact, forest conservation, global warming, wildlife and fish environmental science, Agro forestry, Water resources, community forestry, Silviculture, Remote sensing, GIS, environmental science management, International forestry, soils, recreation, wilderness, management and pathology.

This encyclopaedia integrates all of the diverse approaches that have gone into the making of the modern field of Environmental science. It is unique among available reference works in that it covers the most recent topics of research. Encyclopaedia of Environmental science is intended for all those interested in this field. It will be of immense value to students of all kinds, especially of Environmental science.

It will also be a major source of reference for professionals and students, who require an authoritative and up-to-date guide to assist them in their work. In preparation of this book, the author has freely consulted large number of books and journals so no authenticity is claimed. Author is especially thankful to M/S Anmol Publications Pvt. Ltd. for shaping this book in its final form. Suggestions for further improvement of this book are not only welcome but also greatly appreciated.

Author

A

Abiotic Non-living parts of the environment, e.g., air, water, coal, geological structures, temperature, topography.

Absolute humidity In a system of moist air, the ratio of the mass of water vapour present to the volume occupied by the mixture; that is, the density of the water vapour component. Absolute humidity is normally expressed in grams of water vapour in a cubic meter of air (25g/m^3).

Absorption The process in which radiant energy is retained by a substance. A further process always results from absorption, that is, the irreversible conversion of the absorbed radiation into some other form of energy within and according to the nature of the absorbing medium. The absorbing medium itself may emit radiation, but only after energy conversion has occurred.

Abundance, Relative The frequency of encountering an organism in a given time and place relative to the frequency of encounter in another time or place. Typically the smaller divided by the larger number, thus a proportion.

Abundance, Species An estimate of the total number of individuals of a species in a defined area, population, or community. (Confusion between abundance and density can be avoided if density is viewed as abundance per unit of area or volume).

Access Permission or legal authority to enter an area; entry points or means (e.g., bridges or fords); the mode of travel available within an area. Access into roadless areas is typically provided for hikers, horseback riders, and, in some cases, mountain bikers. Access with motorized vehicles is not provided or allowed in roadless areas.

Acclimation Adjustment to environmental change by an individual. The physiological adjustment or increased tolerance shown by an individual organism to environmental change.

Acid Deposition Precipitation and dry particulate fallout with low pH linked to other components of air pollution.

Acid rain Acids form when certain atmospheric gases (primarily carbon dioxide, sulphur dioxide, and nitrogen oxides) come in contact with water in the atmosphere or on the ground and are chemically converted to acidic substances. Oxidants play a major role in several of these acid-forming processes. Carbon dioxide dissolved in rain is converted to a weak acid (carbonic acid). Other gases. primarily oxides of sulphur and nitrogen, are converted to strong acids (sulphuric and nitric acids). Although rain is naturally slightly acidic because of carbon dioxide, natural emissions of sulphur and nitrogen oxides, and certain organic acids, human activities can make it much more acidic. Occasional pH readings of well below 2.4 (the acidity of vinegar) have been reported in industrialized areas.

The principal natural phenomena that contribute acid-producing gases to the atmosphere are emissions from volcanoes and from biological processes that occur on the land, in wetlands, and in the oceans. The effects of acidic deposits have been detected in glacial ice thousands of years old in remote parts of the globe. Principal human sources are industrial and power-generating plants and transportation vehicles. The gases may be carried hundreds of miles in the atmosphere before they are converted to acids and deposited.

Since the industrial revolution, emissions of sulphur and nitrogen oxides to the atmosphere have increased. Industrial and energy-generating facilities that burn fossil fuels, primarily coal, are the principal sources of increased sulphur oxides. These sources, plus the transportation sector, are the major originators of increased nitrogen oxides.

The problem of acid rain not only has increased with population and industrial growth, it has become more widespread. The use of tall smokestacks to reduce local pollution has contributed to the spread of acid rain by releasing gases into regional atmospheric circulation. The same remote glaciers that provide evidence of natural variability in acidic deposition show in their more recently formed layers, the increased deposition caused by human activity during the past half century.

Acid Soil Soil with a low pH, less than 7.0 but practically less than 6.6.

Acre 43560 square feet of area or 10 units of land 66 feet x 66 feet (a chain is 66 feet).

Acres, Marginal A component of the regulated commercial forest-land area that includes areas not qualifying as standard or special components primarily because of excessive development costs, low product values, resource protection constraints, and inadequate markets.

Acres, Standard Areas where crops of industrial wood can be grown and harvested with adequate protection of forest resources, under the usual provisions of a timber sale contract.

Active system (active sensor) A remote-sensing system that transmits its own radiation to detect an object or area for observation and receives the reflected or transmitted radiation. Radar is an example of an active system. Compare with passive system.

Activity A measure, course of action, or treatment that is undertaken to directly or indirectly produce directly or indirectly, produce, enhance, or maintain forest and rangeland and related rural land outputs or achieve administrative or environmental quality objectives.

Activity Fuels Fuels which have been directly generated or altered by management action. More commonly called slash.

Adaptability The ability to alter structure or function in response to changes in the environment in such a way that fundamental life processes can continue to cope with (or continue to function in) an uncertain environment.

Adaptation The result of the process of 1. Long-term evolutionary adjustment of a population to environmental changes, or.

2. Short-term physiological/behavioural responses compensating for environmental

changes. Frequently a short-term behavioural or physiological response, if it is generally adaptive, will be reinforced by genetic adaptation that ensures the longevity of the short-term response.

Adult Sexually mature; an animal that has contributed new individuals to a population.

Advance Reproduction/ Advance Regeneration True seedlings or sprouts of the tree species that occupy the overstory existing as part of the understory vegetation. The root systems of sprout advance reproduction are usually older than the stems, are relatively large, and are capable of supporting much greater stem growth than the root stems of true seedlings.

Advanced Very High Resolution Radiometer (AVHRR) A five-channel scanning instrument that quantitatively measures electromagnetic radiation, flown on NOAA environmental satellites. AVHRR remotely determines cloud cover and surface temperature. Visible and infrared detectors observe vegetation, clouds, lakes, shorelines, snow, and ice. TIROS Automatic Picture Transmissions (APT) are derived from this instrument.

Adventitious Bud A bud from a tree bole or stem usually resulting from a branch being broken or the bole being exposed to sunlight.

Adverse Impact An effect that causes harmful interference (beyond limits of acceptable change) with the structure or functioning of a person, place, ecosystem, or object.

Aerosol Particules of liquid or solid dispersed as a suspension in gas.

Aesthetics Relating to or dealing with perceptions of beauty; the philosophy dealing with beauty and with judgments or perception of beauty; also esthetic.

Afforestation The act or process of establishing a forest, especially on land not previously forested.

Age Class A general term applied to organisms which are born, hatched, etc. within a specified time period; also a life stage. Age class time frame varies with generation time of organisms. For example, the young, juvenile, and adult age classes of many small mammals occur within a single year, whereas for trees, the young age class encompasses several years, typically 10. Age intervals into which wildlife or timber stands are divided are commonly 10 years. Age classes are "balanced" when the whole forest has an approximate equal number of acres in each 10-year class from age zero through rotation age.

Age Composition 1. An array of age classes.

2. The arrangement of age classes in a population which describes the relative strengths of the age classes. Synonymous with *age structure.*

Age Distribution (Age Class Distribution) Individuals of a population classified according to age classes or periods such as pre-reproductive or sub-adult, reproductive, and post-reproductive. An *age pyramid* is a diagram of the proportions within each age class of a population.

Age Ratio Proportion of young to adults, suggesting a measure of productivity.

Air mass Large body of air often hundreds or thousands of miles across, containing air of a similar temperature and humidity. Sometimes the differences between air masses are hardly noticeable, but if colliding air masses have very different temperatures and humidity values, storms can erupt.

Air pollution The existence in the air of substances in concentrations that are determined unacceptable to human health and the environment. Contaminants in the air we breathe come mainly from manufacturing industries, electric power plants, exhaust from automobiles, buses, and trucks.

Air pressure The weight of the atmosphere over a particular point, also called barometric pressure. Average air exerts approximately 14.7 pounds (6.8 kg) of force on every square inch (or 101,325 Newton on every square meter) at sea level.

Albedo The ratio of the outgoing solar radiation reflected by an object to the incoming solar radiation incident upon it.

Algae Simple rootless plants growing in water and in soil. Algal blooms may adversely affect water quality.

Algorithm A mathematical relation between an observed quantity and a variable used in a step-by-step mathematical process to calculate a quantity.

In the context of remote sensing, algorithms generally specify how to determine higher-level data products from lower-level source data. For example, algorithms prescribe how atmospheric temperature and moisture profiles are determined from a set of radiation observations originally sensed by satellite sounding instruments.

Alkaline Substance capable of neutralizing acid, with a pH greater than 7.0.

All-aged Forest A forest which contains trees of all ages, intermixed.

Allowable Cut The amount of forest wood (or forest products) that may be harvested annually or periodically from a specified area over a stated period in accordance with the objectives of management.

All-Terrain Vehicle (ATV) A type of off-highway motorized vehicle. These include vehicles with low-pressure tires, commonly called three-wheelers and four-wheelers. Because all-terrain vehicles cannot be licensed or registered under state vehicle laws, they may not be allowed on certain types of roads.

Alluvial Fan A fan-shaped deposit of soil and rock formed by a stream where the velocity is abruptly decreased, as at the mouth of a ravine or at the foot of a mountain.

Alluvium Material such as earth, sand, gravel, or other rock or mineral materials transported and laid down by flowing water.

Alpha Diversity Any of a variety of definitions of diversity relating to lists of species, abundance of each, and proportional abundance within a community. It is an expression of populations co-occurring at a specific time and space.

Altimeter An active instrument used to measure the altitude of an object above a fixed level. For example, a laser altimeter can measure height from a spacecraft to an icesheet. That measurement, coupled with radial orbit knowledge, will enable determination of the topography.

Altricial Animals which, after birth or hatching, delay attaining a condition of self maintenance.

Amplitude The magnitude of the displacement of a wave from a mean value. For a simple harmonic wave, it is the maximum displacement from the mean. For more complex wave motion, amplitude is usually taken as one-half of the mean distance (or difference) between maxima and minima.

Amplitude, Ecological The range of one or more environmental conditions in which an organism or a process can function.

Analgesia The reduction or removal of the pain sense without loss of conciousness following drug administration. Narcosis is a condition of analgesia accompanied by deep sleep or stupor.

Analogue Transmission of a continuously variable signal as opposed to a discretely variable signal. Compare with digital. A system of transmitting and receiving information in which one value (i.e., voltage, current, resistance, or, in the APT system, the volume level of the video tone) can be compared directly to the information (in the APT system, the white, black, and gray values) in the image.

Analysis Area A delineated area of land subject to analyses of 1. Responses to proposed management practices in producing, enhancing, or maintaining forest and rangeland outputs and environmental quality objectives, and.

2. Economic and social impacts.

Anemometer Instrument used to measure wind speed usually measured either from the rotation of wind driven cups or from wind pressure through a tube pointed into the wind.

Angle of Repose The maximum angle (or percent slope) to the horizontal at which any soil or loose or fragmented solid material will stand without sliding, e.g., soil on a roadside cut.

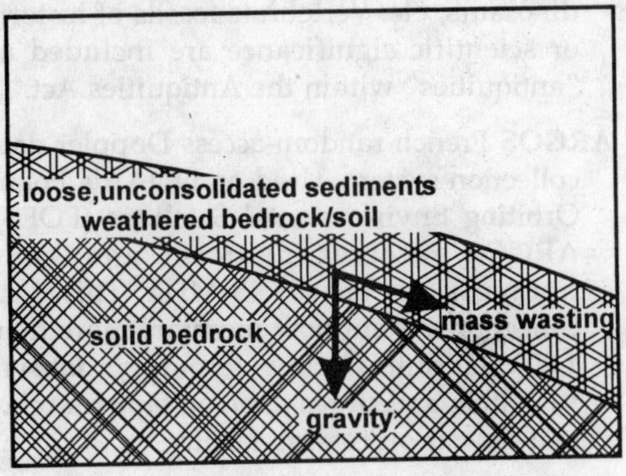

Fig. Angle of Repose

Angle of Slide The slope, measured in degrees of deviation from the horizontal, on which loose or fragmented materials will start to slide, slightly greater than the angle of repose.

Animal Community The characteristic assemblage of animals associated with a.

Animal Unit A measure for converting types of animals to forage resources. It is based on the equivalent of the forage required by a mature cow of about 1000 pounds.

Animal, Wild 1. Animals, or their progeny, which once were domesticated but escaped captivity and are running wild (i.e., feral animals). Horses, burros, and hogs are not considered to be wildlife.

2. Undomesticated vertebrate animals except fish, considered collectively.

3. Generally, all non-domesticated non-plant life including microscopic forms, insects, crustaceans, mollusks, and fish.

4. More particularly, a loose term that includes non-domesticated vertebrates, especially mammals, birds, and fish, and some of the higher invertebrates (such as crabs, crayfish, etc.).

5. Living things that are neither plant, human, nor domesticated; especially the mammals, birds, and fishes that are hunted by people for sport or food.

Annuals Plants that normally live less that one year.

Anomaly 1. The deviation of (usually) temperature or precipitation in a given region over a specified period from the normal value for the same region.

2. The angular distance of an Earth satellite (or planet) from its perigee (or perihelion) as seen from the center of the Earth (sun).

Antagonism The combined actions of two or more agents, the result of which is less than the sum of the actions of each of the agents acting separately or independently. Antagonism may be "results" or merely the net "actions" themselves.

Antecedent Moisture (Antecedent Soil Water) The degree of wetness of the soil at the beginning of a runoff period expressed as an index or as the total inch-depth-equivalents of water stored in the soil.

Anthracnose A disease characterized by ulcer-like leaf or fruit spots and caused by fungi that produce asexual spores in the type of fruiting body called a acervulus.

Anticyclone A high pressure area where winds blow clockwise in the Northern Hemisphere and counter-clockwise in the Southern Hemisphere.

Apogee On an elliptical orbit path, point at which a satellite is farthest from the Earth.

Aquatic Pertaining to standing and running water in streams, rivers, lakes, ponds, and reservoirs.

Aquifer Layer of water-bearing permeable rock, sand, or gravel capable of providing significant amounts of water.

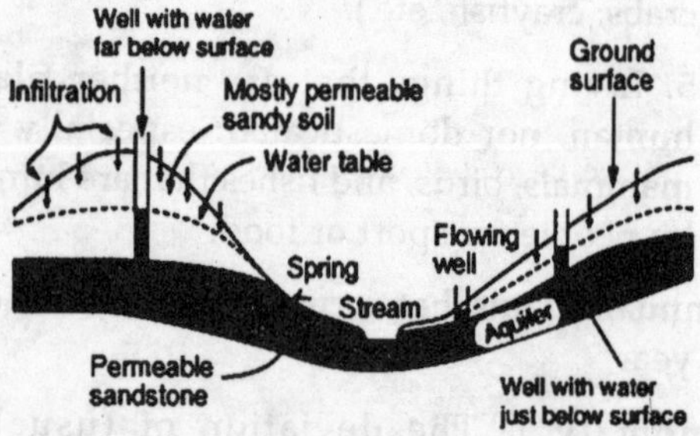

Fig. Aquifer

Archeological Area Sites and areas which have been designated as containing important evidence and remains of the life of people on Earth between the time of their first appearance and the development of written history. Archeological remains are the sole kind of evidence that applies to pre-history and they may provide information about all aspects of human life.

Archeological Resource All remaining physical evidence of former occupation by now-extinct cultural groups. Evidence includes skeletons, settlement remains, implements, artifacts, monuments, and inscriptions. This subclass is distinguished from its more recent counterpart "historical resources" by a point in time between it an movement of settlers into an area. This resource class does not include any of the more recent physical products of contemporary native cultural groups such as may exist on Indian reservations.

Arctic circle The parallel of latitude that is approximately 66.5 degrees north of the equator and that circumscribes the northern frigid zone.

Area Mining Surface mining that is carried on in level to gently rolling topography on relatively large tracts.

Area Reconnaissance Early visits to observe an area and its ecological, economic, and social conditions and to determine the scale and scope of plans and subsequent work (or visits and information collection.).

Area, Geologic Areas containing scientifically or educationally significant fossilized specimens of flora and fauna from previous geologic periods e.g., coal swamp forests, dinosaurs, etc. Vertebrate fossils of historic or scientific significance are included as "antiquities" within the Antiquities Act.

ARGOS French random-access Doppler data collection system. Used on NOAA's Polar-Orbiting Environmental Satellites (POES). ARGOS receives platform and buoy transmissions on 401.65 MHz. This data collection system now monitors more than 4,000 platforms worldwide, outputs data via VHF link, and stores them on tape for relay to a central processing facility.

Arterial Roads Roads and trails that provide service to large land areas and that usually connect with public highways or other forest arterial roads to form an integrated network of primary travel routes. Their location and standard are often determined by a demand for maximum mobility and travel efficiency rather than specific resource management service. Such roads are usually developed and operated for long-term land and resource management purposes and constant service.

Arthropod Invertebrate animals with jointed legs and a segmented body such as insects, mites, and spiders.

Artificial Feeding Artificial feeding of wild animal life means placing food into the environment that supplements the food sources contained naturally within in the home range of a wild species. Usually directed at populations of animals, it may apply to feeding individual animals.

Artificial intelligence Neural networks. The branch of computer science that attempts to program computers to respond as if they were thinking—capable of reasoning, adapting to new situations, and learning new skills. Examples of artificial intelligence programs include those that can locate minerals underground and understand human speech.

Artificial Regeneration or Reproduction Planting or seeding to establish a new crop of trees on an area, usually timber has been removed.

Aspect (Exposure) The compass direction in degrees azimuth toward which a land slope faces (the direction downhill). Since plants suffer their greatest moisture losses through evapotranspiration on southwest facing slopes (which receive the greatest energy input from the sun compared to that received by other slopes), aspect can be an important factor influencing vegetation distribution. Giles (in 1999) recommended separating and studying the influences of north-south aspects (primarily solar influences) and east-west aspects (primarily coastal, anthropogenic, wind and rain influences).

Aspect ratio The ratio of image width to image height. Weather Facsimile images have a 1:1 aspect ratio (square); a conventional TV aspect ratio is 4:3 (rectangle).

Assessment Results of analyzing and interpreting the existing social, economic, or ecological characteristics or conditions of an area using scientific principles (e.g., as they may affect "sustainability").

Assimilation Anabolism or constructive metabolism; the process of taking in and converting nutrient substances into components of living tissues; the conversion of non-living matter into protoplasm.

Association 1. A commonly-occurring relationship between two or more species which can be readily recognized in the field and treated as an ecological unit;.

2. An actual or concrete community, stand, or group of organisms characterized by a definite floristic composition, presenting uniformity in physiognomy and structure, and growing under uniform habitat conditions;.

3. A group of things classified together because they meet certain standards of similarity;.

4. A climatic climax unit that includes all of the successional stages preceding or associated with it. Plant associations and animal associations emphasize populations of plants or animals respectively within an area.

Assumption A judgmental decision by planners which supplies missing values, relationships, or societal preferences for some informational component necessary for making a decision. The true nature of such missing information is either unknown or cannot be readily ascertained within the time and/or budgetary constraints on a decision.

Astronomical Unit (AU) The distance from the Earth to the sun. On average, the sun is 149,599,000 kilometers from Earth.

ATLAS (Atmospheric Laboratory for Applications and Science) Mission The focus of ATLAS is to study the chemistry of the Earth's upper atmosphere (mainly the stratosphere/mesosphere) and the solar radiation incident on the Earth system (both total solar irradiance and spectrally resolved radiance, especially ultraviolet). Science operations onboard ATLAS 1 (March 1992) and ATLAS 2 (March-April, 1993) began a comprehensive and systematic collection of data that will help establish benchmarks for atmospheric conditions and the sun's stability.

Atmosphere The air surrounding the Earth, described as a series of shells or layers of different characteristics. The atmosphere, composed mainly of nitrogen and oxygen with traces of carbon dioxide, water vapour, and other gases, acts as a buffer between Earth and the sun. The layers, troposphere, stratosphere, mesosphere, thermosphere, and the exosphere, vary around the globe and in response to seasonal changes.

Troposphere stems from the Greek word tropos, which means turning or mixing. The troposphere is the lowest layer of the Earth's atmosphere, extending to a height of 8-15 km, depending on latitude. This region, constantly in motion, is the most dense layer of the atmosphere and the region that essentially contains all of Earth's weather. Molecules of nitrogen and oxygen compose the bulk of the troposphere.

The tropopause marks the limit of the troposphere and the beginning of the stratosphere. The temperature above the tropopause increases slowly with height up to about 50 km.

The stratosphere and stratopause stretch above the troposphere to a height of 50 km. It is a region of intense interactions among radiative, dynamical, and chemical processes, in which horizontal mixing of gaseous components proceeds much more rapidly than vertical mixing. The stratosphere is warmer than the upper troposphere, primarily because of a stratospheric ozone layer that absorbs solar ultraviolet energy.

The mesosphere, 50 to 80 km above the Earth, has diminished ozone concentration and radiative cooling becomes relatively more important. The temperature begins to decline again (as it does in the troposphere) with altitude. Temperatures in the upper mesosphere fall to -70o to -140oCelsius, depending upon latitude and season. Millions of meteors burn up daily in the mesosphere as a result of collisions with some of the billions of gas particles contained in that layer The collisions create enough heat to burn the falling objects long before they reach the ground.

The stratosphere and mesosphere are referred to as the middle atmosphere. The mesopause, at an altitude of about 80 km, separates the mesosphere from the thermosphere—the outermost layer of the Earth's atmosphere.

The thermosphere, from the Greek thermo for heat, begins about 80 km above the Earth. At these high altitudes, the residual atmospheric gases sort into strata according to molecular mass. Thermospheric temperatures increase with altitude due to absorption of highly energetic solar radiation by the small amount of residual oxygen still present. Temperatures can rise to 2,000oC. Radiation causes the scattered air particles in this layer to become charged electrically, enabling radio waves to bounce off and be received beyond the horizon.

At the exosphere, beginning at 500 to 1,000 km above the Earth's surface, the atmosphere blends into space. The few particles of gas here can reach 4,500oF (2,500oC) during the day.

Atmospheric pressure The amount of force exerted over a surface area, caused by the weight of air molecules above it. As elevation increases, fewer air molecules are present. Therefore, atmospheric pressure always decreases with increasing height. A column of air; 1 square inch in cross section, measured from sea level to the top of the atmosphere would weigh approximately 14.7 lb/in2. The standard value for atmospheric pressure at sea level is:.

29.92 inches or 760 mm of mercury.

1013.25 millibars(mb) or 101,325 pascals (Pa).

Atmospheric Radiation Measurements Program (ARM) U.S. Department of Energy program for the continual, ground-based measurements of atmospheric and meteorological parameters over approxi-mately a ten-year period. The program will study radiative forcing and feedbacks, particularly the role of clouds. The general program goal is to improve the performance of climate models, particularly general circulation models of the atmosphere.

Atmospheric response variables Variables that reflect the response of the atmosphere to external forcing (e.g., temperature, pressure, circulation, and precipitation).

Atmospheric windows The range of wavelengths at which water vapour; carbon dioxide, or other atmospheric gases only slightly absorb radiation. Atmospheric windows allow the Earth's radiation to escape into space unless clouds absorb the radiation. See greenhouse effect.

Atoll A coral island consisting of a ring of coral surrounding a central lagoon. Atolls are common in the Indian and Pacific Oceans.

Attitude A learned predisposition, feeling, or conviction manifesting itself in a general state of readiness either to evaluate or to react toward an object or class of objects in either a favourable or unfavourable manner in a more or less consistent and characteristic way. Attitudes are relatively stable and have three components: a cognitive or belief component, an affective or feeling component, and a connotative or action-disposition component. Although behaviour may be a function of attitude at times, attitudes are not necessarily predictors of behaviour. Consequently, changes in attitudes do not always result in changes in behaviour.

Attribute A characteristic of a component. The sum of knowledge of all attributes defines a component, e.g., attributes of grass include colour, phenology, live biomass, dead biomass, litter biomass, etc.

Audio frequencies Frequencies that the human ear can hear (usually 30 to 20,000 cycles per second).

Auger Mining Generally practiced but not restricted to hilly coal-bearing regions of the country. Utilizes a machine designed on the principle of the storage pile or bin for loading and transporting. May be used alone or in combination with conventional surface mining. When used alone, a single cut is made sufficient to expose the coal seam and to provide operating space for the machine. When used in combination with surface mining, the last cut pit provides the operating space.

Automatic Picture Transmission (APT) System developed to make real-time reception of satellite images possible whenever an APT-equipped satellite passes within range of an environmental satellite ground station. Transmission (analogue video format) consists of an amplitude-modulated audible tone that can be displayed as an image on a computer monitor when received by an appropriate ground station.

APT images are transmitted by polar-orbiting satellites such as the TIROS-N/NOAA satellites and Russia's METEOR, which orbit 500-900 miles above the Earth, and offer both visible and infrared images. An APT image has thousands of squares called picture elements or pixels. Each pixel represents a four-km square.

Autotrophic Organism An organism capable of synthesizing organic matter from inorganic substances. In contrast, a heterotrophic organism must have organic matter as its source of food.

Average Daily Traffic (ADT) The average 24-hour volume of traffic, being the total volume of traffic during a stated period divided by the number of days in that period. Unless otherwise stated, the period is 1 year.

Azimuth Compass direction from a point, measured in degrees clockwise from true North.

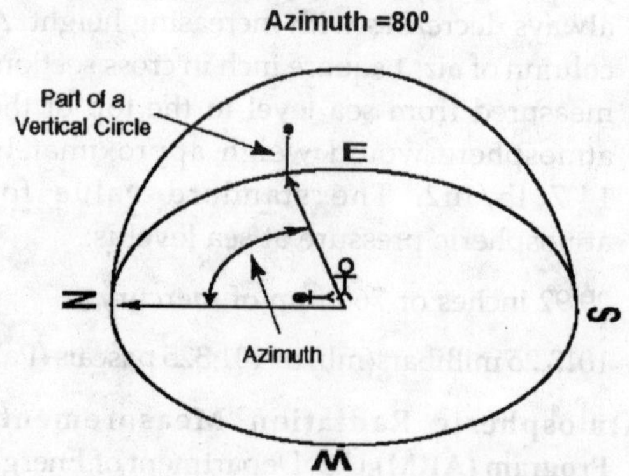

Fig. Azimuth

Azimuth The direction, in degrees referenced to true north, that an antenna must be pointed to receive a satellite signal (compass direction). The angular distance is measured in a clockwise direction.

B

Backcountry An imprecise term often used to refer to all those portions of wildlands in which usually there are no permanent, improved, or maintained access roads or operational, fixed facilities (such as lumber mills, ski resorts, or settlements with permanent residents). Deteriorating, unused and unoccupied structures may be present. Those roads which are present are only usable by trucks or 4 wheeled-drive vehicles and typically dead-end in the "back country", rather than continue through it. Any presently active uses only have primitive facilities, e.g., cabins, base camps, undeveloped campgrounds. In common usage, primitive and wilderness areas are backcountry types of wildlands which show no obvious evidence of permanent human uses or occupancy. These are any areas where the management objectives stress dispersed, off-road recreation activities, e.g., hiking, trail bike riding, hunting, and fishing. "Backcountry" is a recreation area while "wilderness" is defined primarily as a large natural ecosystem, to be experienced as it is. Recreational opportunities with structural elements can be enhanced or even created in "backcountry", but rarely in "wilderness".

Backcountry Recreation A near-primitive recreation experience within the natural setting of a relatively roadless area where travel is primarily by horse or foot trails or by canoe.

Backfill The operation of filling an excavation or adding height to a road. Also the material placed in an excavation in the process of backfilling.

Backfire 1. Fire skillfully set along the inner edge of a fire control line to stop a spreading wildfire by reducing the fuel or changing the direction of force of the convection current of the fire. Using such fire to consume unburned fuel inside the fireline to speed up the line holding and mopup is usually called "burning out" or "clean burning.".

2. A prescribed fire set to burn against the wind;.

3. To set a backfire.

Bajada An alluvial plain formed at the base of a mountain by the joining of several alluvial fans.

Balance of Nature (Ecological Stability) The apparent stability of the population density relationships between the many species of organisms that make up a biotic community. The concept of such a balance has been severely criticized. Population density of each of the species in a biotic community may fluctuate more or less widely from time to time (e.g., existence of population explosions, die-offs, epidemics, irregular migrations, succession etc.) and that the density of no species remains at a fixed value. The fluctuations in the numbers of each species, however, usually have fairly definite limits. Using 'balance' includes the recognition that the population densities of every species making up a community fluctuate from season to season and from year to year.

Band 1. In radio a continuous sequence of broadcasting frequencies within given limits.

2. In radiometry, a relatively narrow region of the electromagnetic spectrum to which a remote sensor responds; a multispectral sensor makes measurements in a number of spectral bands.

3. In spectroscopy spectral regions where atmospheric gases absorb (and emit) radiation, e.g., the 15 μm carbon dioxide absorption band, the 6.3 μm water vapour absorption band, and the 9.6 μm ozone absorption band.

Bare root Seedling Young trees shipped without their roots being in soil.

Barometer An instrument used to measure atmospheric pressure. A standard mercury barometer has a glass column about 30 inches long, closed at one end, with a mercury-filled reservoir. Mercury in the tube adjusts until the weight of the mercury column balances the atmospheric force exerted on the reservoir. High atmospheric pressure forces the mercury higher in the column. Low pressure allows the mercury to drop to a lower level in the column. An aneroid barometer uses a small, flexible metal box called an aneroid cell. The box is tightly sealed after some of the air is removed, so that small changes in external air pressure cause the cell to expand or contract.

Basal Area (BA) 1. The area in square feet of the cross section of a tree including bark, at 4.5 feet above the ground. A tree 14 DBH is approximately 1 square foot. Generally expressed as the total of all trees, thus total Basal Area per acre. Measurement of how much of a site is occupied by trees e.g., 90-100 square feet per acre.

2. In range management, it is the area of ground surface occupied by the stem(s) of a range plant, as contrasted with the full spread of its herbage or foliage, generally measured at one inch above soil level.

Base A substance that forms a salt when it reacts with acid. A base is a substance that removes hydrogen ions (protons) from an acid and combines with them in a chemical reaction.

Base Flow That portion of the water flowing in a stream which is due to ground water seepage into the channel. Often the minimum length of flowing water in a summer drought.

Base Map A map showing certain fundamental information, on which is compiled additional data of specialized nature. Also a map to which all other maps and GIS map layers are registered and re-scaled.

Baseline Study, Ecological 1. For a geographic area where human activities potentially affecting the environment are proposed, a study is undertaken to describe the existing conditions and trends. This description of key ecological system components and their functional relations through time provide a reference "baseline" from which the effects of the proposed action may be predicted. It also provides a benchmark against which future conditions and trends, occurring after the proposed activity begins, may be compared. Thus the purpose of such a study is to provide sound ecological information as input to the decision-making process for locating, designing, and operating new human activities which are potentially degrading to the ecological environment.

Baud Unit of signaling speed. The speed in bauds is the number of discrete conditions or signal events per second. If each signal event represents only one bit condition, baud is the same as bits per second.

Bay A wide area of water extending into land from a sea or lake.

The measure of the "width" of an antenna pattern, measured in degrees of arc. Generally an antenna with low gain has a wide pattern, receiving signals well from a number of different directions.

Bed Load Soil, rock particles, or other debris rolled along the bottom of a stream by the moving water, as contrasted with the "silt load" carried in suspension.

Bedding Area A specific site selected by big game animals to lie down and rest.

Bedrock The more or less solid rock in place either on or beneath the surface of the Earth. It may be soft or hard and have a smooth or irregular surface.

Beginners All-purpose Symbolic Instruction Code (BASIC) A most popular and widespread "high level" language for microcomputers. BASIC uses a sequence of English-like commands and statements.

Belief An unproven assertion based on one or more fundamental assumptions. The assertion may be unprovable.

Bench The surface of an excavated area at some point between the material being mined and the original surface of the ground on which equipment can sit, move, or operate. A working road or base which is below a highwall as found in contour stripping for coal.

Benchmark 1. Sites for measuring natural phenomena in areas where human activity has essentially no effect. These sites would be expected to continue at least as long as natural conditions exist.

2. Certain existing examples or conditions in the field which are usable as established standards or definitions for practical comparison of "use capability" classes. Units of equal "use capability", but possessing markedly different physical features or appearances, can be recognized as being equal.

3. A set of estimates used to establish standards by which to compare alternatives considered in detail. Benchmark alternatives include minimum level, maximum resource levels, and maximum present net value levels.

Benefit (Outcome) Inclusive terms used to quantify the positive expected results or outputs of a proposed activity, project, or program expressed in monetary or non-monetary terms. Ideally, estimates of all benefits, outputs, or effectiveness expected to be received or achieved as a result of undertaking a proposed activity, program, or project. Realistically, a particular program and its outputs, some intended and very specific beneficial outputs, although some attendant outputs may be negative or non-beneficial.

Benthos The whole group of organisms in a water body (pond, lake, or river) which crawl about on the bottom, or burrow into the bottom, or grow attached to the bottom. Those aquatic organisms that are carried passively by water currents or the wind are termed plankton and those that swim actively and may move long distances for feeding or breeding purposes, nekton.

Berm 1. A strip of coal left in place temporarily for use in hauling or stripping. A layer of large rock or other relatively heavy stable material placed at the outside bottom of the spoil pile to help hold the pile in position (a toe wall). Also used higher in the spoils for the same purpose.

2. A small mound of earth on the outer edge of a mountain or secondary road.

Beta Diversity Any of a variety of definitions of diversity relating to differences among communities present along a gradient (e.g., soil acidity) or in the same place over time.

Biltmore Stick Like a yard stick, calibrated to measure the diameter of a tree at breast height.

Binary A numbering system that uses only 1 and 0 (e.g., 1 is one, 10 is two, 11 is three). In digital integrated circuits, a 0 is indicated by a logic low and a 1 by a logic high.

Bio(geo)c(o)enose The translated Germanic and Slavic language equivalent of ecosystem.

Bioaccumulation The uptake and retention of environmental substances by an organism by means other than the ingestion of food. See Biomagnification.

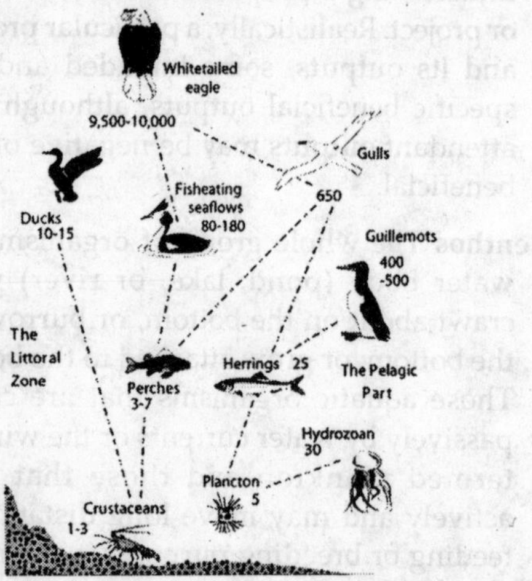

Fig. Bioaccumulation

Bioassay A measurement of the effects of a substance on living organisms.

Biochemical Oxygen Demand (also BOD; Biological Oxygen Demand) A measure of the quantity of dissolved oxygen, in milligrams per litre, required to stabilize the demand for oxygen in a water sample, usually resulting from the process of microorganisms consuming organic matter and utilizing the available dissolved oxygen in the oxidation process. Adding waste to water may cause a high BOD, deplete oxygen, and kill aquatic life.

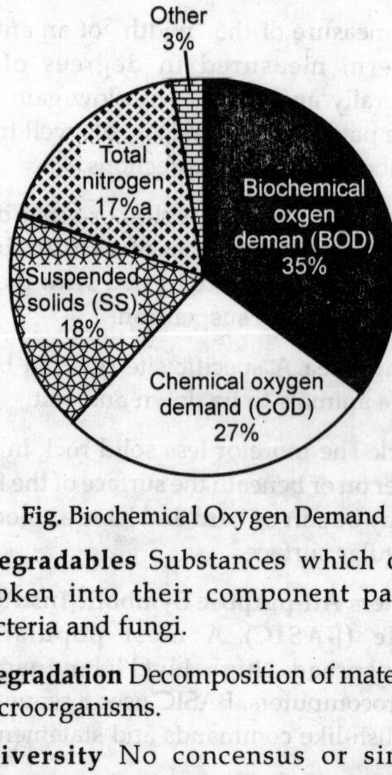

Fig. Biochemical Oxygen Demand

Biodegradables Substances which can be broken into their component parts by bacteria and fungi.

Biodegradation Decomposition of material by microorganisms.

Biodiversity No concensus or singular definition exists but it is included within more than 20 laws and regulations. It has various denotations and connotations. Its use is usually limited to animals, though plants and intermediate forms are "life". Microscopic forms are usually excluded and insects are rarely included. Distributions of faunalmass and phytomass have been implied by some workers. Most texts use species richness (the count and list of species) as a synonym. Some use indices of proportionate occurrence or rareness (Shannon and related indices) as a specific meaning. Whether constant abundance of plants (or animals) over many years (great evenness) is desirable or required by the definition is debated. Rapidly changing modern taxonomy, inclusion of migrants, sampling intensities, descriptions of patterns occupied in nature, appropriate levels of statistical confidence, and even the categories of life to be included are also debated and thus collectively result in a general word almost without meaning.

Biogeochemical Relating to the chemical relationships between the geology of an area and its plant and animal life. In land use planning, the term is more often used for all of the naturally-occurring objects, processes, and relationships in an area. Often used in the sense of the biogeochemical parameters of a planning area. *Biogeochemical* is approximately equivalent to the natural ecosystem and ecosystem functions of a planning area. This composite term is based on the assumption that all naturally-occurring things can be classified as being biological, geological, or chemical.

Biogeochemical Cycle (Mineral Cycle, Nutrient Cycle, Material Cycle) Movements

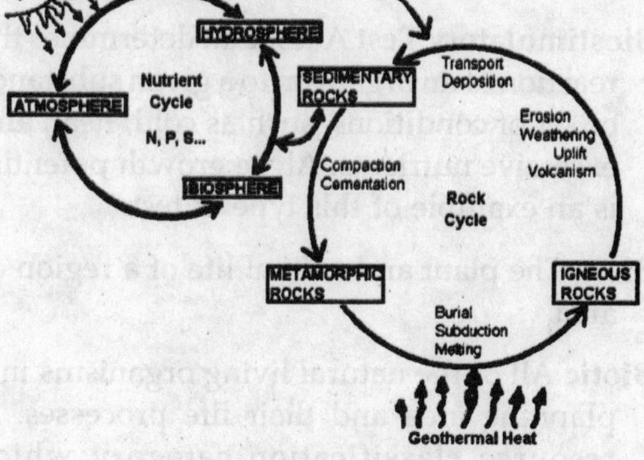

Fig. Biogeochemical Cycle

through the Earth system of key chemical constituents essential to life, such as carbon, nitrogen, oxygen, and phosphorus. The circulation (cycling) of chemical elements such as nitrogen, carbon, etc. in specific pathways from the abiotic portions of the environment into organic substances (flora and fauna) and then back again into abiotic forms.

Biogeophysical A composite term often used to encompass all of the naturally occurring materials, processes, and relationships operating in an area. Often used in the sense of the "biogeophysical parameters" of a planning area. In such usage, the term is approximately equivalent to the natural ecosystem and ecosystem functions of a planning area. The word was constructed so that all the natural objects and processes operating in an area can be classified as being either biological (i.e., plant and animal species, ecological interactions, biotic productivity, etc.) or geological (i.e., rock types, soil types, geomorphic history, sedimentation, erosion, etc.), or physical (i.e., heat, light, electrical, gravitational, etc.).

Biologic Diversity (Species Diversity, Biotic Diversity, Biodiversity) Usually equivalent to species richness, i.e., the number of different species occurring in some location or under some condition such as pollution. "Biologic diversity" may also be used in a more general sense to refer to the number of higher taxonomic levels or types and amounts of organismal relationships in some location or under some condition — e.g., the number of genera, families, orders or phyla present; or genetic pathways; or the number of biotic communities present; or the number of energy, nutrient, or food-chain pathways present. There are many measures and indices. Frequently used relationships are those for proportions of plants, animals, or areas present.

Biological Control Using organisms or viruses to control parasites, weeds, or pests.

Biological Growth Potential The average net growth attainable in a fully stocked area of forest land.

Biomagnification (Bioconcentration) The accumulation of toxic materials then passage to feeders in food chains. See Bioaccumulation.

Biomass (Protoplasm Mass) 1. The amount of living material in unit area or volume, usually expressed as mass or weight. The total weight (at a given time) of living organisms of one or more species per unit of space (species biomass), or of all the species in a community (community biomass). *Phytomass* expresses weight of plants only.

2. The total weight of matter incorporated into all organisms (living and dead).

Biome Well-defined terrestrial environment (e.g., desert, tundra, or tropical forest). The complex of living organisms found in an ecological region.

Biometry The study of the statistics of biological phenomena; the application of mathematics and statistics to the study of living things.

Biophysical All the naturally occurring objects and processes of an area on the assumption that all naturally occurring things can be classified as being either living (i.e., biotic) or nonliving (physical or abiotic).

Biophysical Determinism A point of view in planning which advocates making resource allocation and use decisions primarily on the basis of the inherent ability of the resource to sustain use, i.e., an ecological carrying capacity concept of resource capability. Economic and social system needs and desires are viewed as being completely subordinated to this inherent capacity of the resource to meet any and all demands while maintaining its natural integrity, condition, and performance of other natural processes at a desirable level. Economic and social demands can only be satisfied within the ecological carrying capacity constraints. The terms "biological primacy," "physiographic determinism," and "ecological determinism," in their common usage, are equivalent to biophysical determinism - though "biophysical" more clearly implies concern with both physical and biological processes.

Biosphere Part of the Earth system in which life can exist, between the outer portion of the geosphere and the inner portion of the atmosphere.

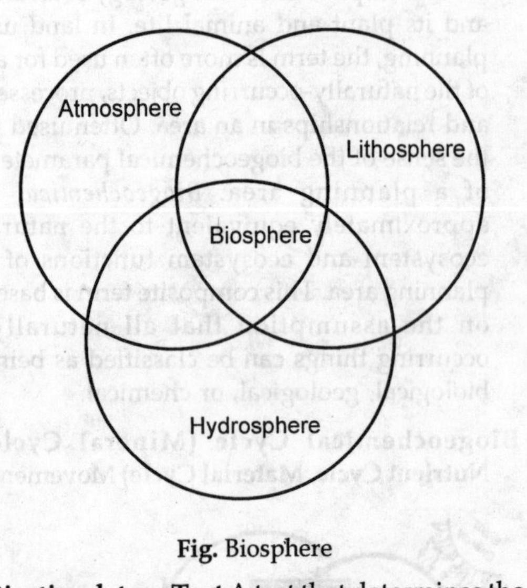

Fig. Biosphere

Biostimulatory Test A test that determines the reaction of an organism to a given substance or set of conditions, such as cold, heat, and excessive nutrients. Algae growth potential is an example of this type of test.

Biota The plant and animal life of a region or area.

Biotic All of the natural living organisms in a planning area and their life processes. A resource classification category which subdivides the natural resources and properties into either the biotic or the abiotic (nonliving) entities and characteristics.

Human activities are not usually grouped with either of these categories because they are not considered to be a part of a planning area's natural entities or characteristics. Nevertheless, native cultures have been actively influencing the course of biotic activity in some areas for centuries.

Biotic Potential The inherent ability of members of a population to grow in numbers within a given time and under stated environmental conditions. This potential depends on the number of live, fertile offspring produced at each reproduction and, where sex is involved, the sex ratio, as well as the age classes distribution of the population.

Biotype A genetically homogeneous population composed only of closely similar individuals; a genotypic race or group of organisms.

Bit A contraction of "binary digit." The basic element of a two-element (binary) computer language.

Black Box An unknown and often unknowable mechanism, process, or system which is judged solely by observing its inputs and outputs.

Blaze To mark a tree, usually with paint or cutting into the bark. Often used along boundaries.

Blizzard A severe weather condition characterized by low temperatures and strong winds (greater than 35 mph) bearing a great amount of snow, either falling or blowing. When these conditions persist after snow has stopped falling, it is called a ground blizzard.

Bloom (Algal Bloom) A readily visible, concentrated, distinctively coloured growth or aggregation of plankton (plant and/or animal) in water.

Bole Main trunk of a tree.

Borrowpit A depression from which soil or rocks are taken to build roads, usually along side the road.

Botanical Area An area which has been designated as containing specimens or group exhibits of plants, plant groups, and plant communities which are significant because of form, colour, occurrences, habitat, location, life history, arrangement, ecology, environment, rarity, and other features.

Boundaries Lines indicating the limits of countries, states, or other political jurisdictions, or different air masses.

Box Cut The initial cut into the earth (as in coal mining) driven in a property where no open side exists; this results in highwalls on both sides of the cut.

Breeding rate Natural inherent capacity to reproduce; numbers per unit time, usually expressed as a theoretical or attained rate.

British Thermal Unit The amount of heat needed to raise the temperature of one pound of water by one degree Fahrenheit. Compare with calorie.

Broadbased Dip A surface drainage structure in a dirt road designed to move water off the road while vehicles maintain normal driving speed.

Broadcast Seeding Scattering seed on the surface of the soil. Contrast with drill seeding, whereby the seed is placed in rows in the soil.

Broadscale Assessment A synthesis of current scientific knowledge, including a description of uncertainties and assumptions, to provide a characterization and comprehensive description of ecological, social, and economic components within an assessment area critical for understanding past and present conditions and projecting fliture trends which provides a foundation for the identification of additional or necessary information for policy discussions or decisions.

Brood Rearing Area "Bugging areas" or habitat providing abundant insects for young turkeys, grouse, or quail and other birds.

Browse 1.That part of the current leaf and twig growth of potentially palatable shrubs, woody vines, and trees available for animal consumption. It includes leaves. Bark of some trees is included. Fallen leaves of some species are eaten and represent an undecided element of the definition.

2. The acts of consuming such growth.

Browse Line 1.The uppermost limit on trees and tall shrubs to which livestock and big game browse. (Once called the plimsole line (seen on the side of ocean vessels)). Also grazing line.

2. The height to which animals have grazed or browsed and removed vegetation over an unspecified time.

Brush Growth (stands or clumps) of shrubs or small trees usually of a type undesirable as forage for livestock or in timber management but of some potential value to wildlife and as cover for livestock.

Brush Control Reduction of brush (volume, height, density, and biomass) to reduce its competition for space, moisture, light, and nutrients with plant species that are preferred for growing timber or forage, or, to reduce wildfire fuel.

Buffer (Buffer Zone) A designated land or water area, along the perimeter of some land use, the use of which is regulated so as to resist, absorb, or otherwise preclude unwanted development or other intrusions (visual, noise, recreational use) into areas beyond the strip of undisturbed vegetation that retards the flow of runoff water, causing deposition of transported material and thereby reducing sediment in receiving streams.

Buffer Species A non-game or comparatively undesirable animal species that is food for predators. Its consumption reduces predation (substituting for it) upon game or other desirable species. Managers attempt to increase buffer species to reduce predation harmful to their species objectives.

Bulk Density The mass or weight of oven-dry soil per unit of bulk volume, including air space. Typically grams per cubic centimeter, water having the weight of one gram per cc. Compaction from logging equipment, recreationists' use, grazing animals, etc. increases bulk density.

Burn 1.Area over which fire has recently run;.

2. A management technique.

Burning Prescription Written direction stipulating fire environment conditions, techniques, and administrative constraints necessary to achieve specified resource management objectives by using fire on a given area of land.

Bus The basic frame of a satellite system that includes the propulsion and stabilization systems, but not the instruments or data systems.

Butt Log The first log above the stump and generally the most valuable log in a tree.

Byte A unit of eight bits of data or memory in microcomputer systems.

C

Cache, Firetool A supply of fire-fighting tools and equipment assembled in planned quantities or standard units at a strategic point for exclusive use in fire supression.

Calibration Act of comparing an instrument's measuring accuracy to a known standard.

Calipers, Tree Used for measuring tree or log diameter.

Calorie The amount of heat needed to raise the temperature of one gram of water at 15ocentigrade by one degree centigrade.

Cambium The layer of cells between the innerbark and wood of a tree (where growth takes place).

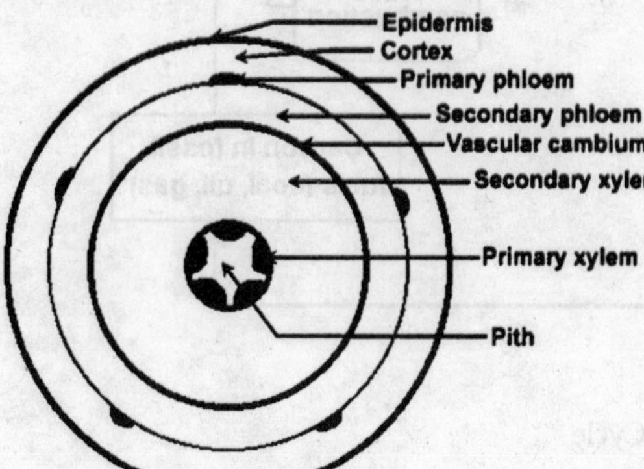

Fig. Cambium

Canal A man-made watercourse designed to carry goods or water.

Candle (tree) The new bright green and tender growth of conifers in the spring.

Canopy The layer formed naturally by the leaves and branches of trees and plants.

Canyon A large but narrow gorge with deep sides.

Capability (Inherent Capability, Natural Capability, Natural Carrying Capacity) Capability refers to evaluations for usability based on the natural, intrinsic, or present state of the resource. The potential of an area of land to produce resources, supply goods and services, and allow resource uses under an assumed set of management practices and at a given level of management intensity. Capability depends on current conditions and site conditions such as climate, slope, landform, soils and geology, as well as the application of management practices, such as silviculture or protection from fires, insects, and disease. Suitability is used for evaluations based on assumptions about potential usability or productivity if specified management alterations were to be made – such as drainage improvements, added irrigation and/or fertilization. Modifiers that make clear what type evaluation procedure is implied by the rating should accompany both terms. "Capability" should always be presented as "inherent capability", "intrinsic capability", or "natural capability". Similarly "suitability" could be referred to as "managed suitability".

Cape (or point) A piece of land extending into water.

Carbon A nonmetallic chemical element occurring in many inorganic compounds and all organic compounds. Diamonds and graphite are pure carbon; carbon is also present, with other substances, in air, coal, and charcoal.

Carbon cycle All parts (reservoirs) and fluxes of carbon. The cycle is usually thought of as four main reservoirs of carbon interconnected by pathways of exchange. The reservoirs are the atmosphere, terrestrial biosphere (usually includes freshwater systems), oceans, and sediments (includes fossil fuels). The annual movements of carbon, the carbon exchanges between reservoirs, occur because of various chemical, physical, geological, and biological processes. The ocean contains the largest pool of carbon near the surface of the Earth, but most of that pool is not involved with rapid exchange with the atmosphere.

Carbon dioxide (CO_2) A minor but very important component of the atmosphere, carbon dioxide traps infrared radiation. Atmospheric CO_2 has increased about 25 percent since the early 1800s, with an estimated increase of 10 percent since 1958 (burning fossil fuels is the leading cause of increased CO_2, deforestation the second major cause). The increased amounts of CO_2 in the atmosphere enhance the greenhouse effect, blocking heat from escaping into

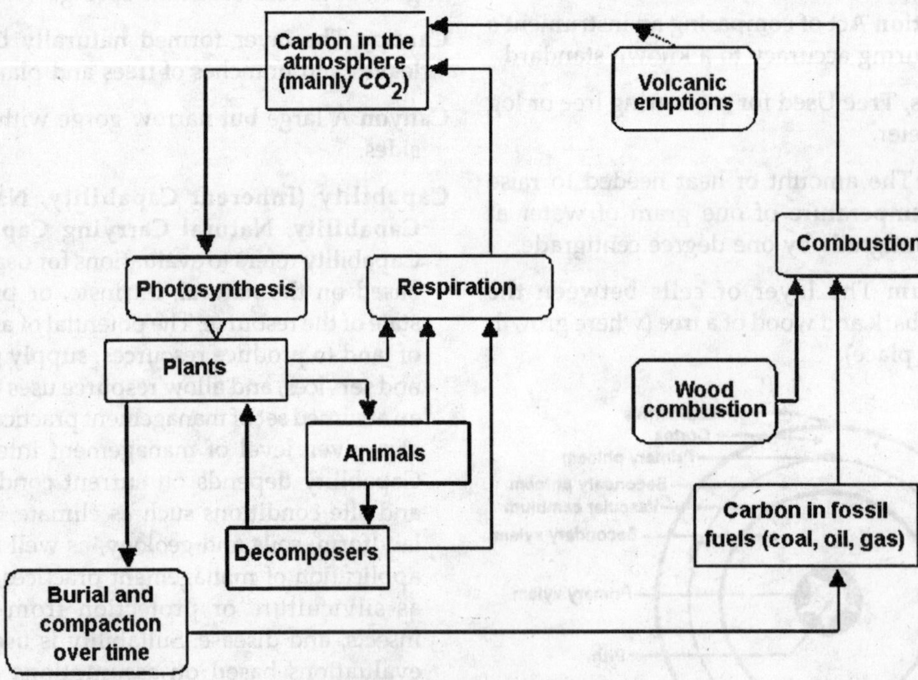

Fig. Carbon Cycle

space and contributing to the warming of Earth's lower atmosphere.

Cardinal Value Numerical value assigned to a variable that relates directly to some physical property (height, weight, etc.). Each number has some meaning by itself about the measured property.

Carnivore 1. A flesh-eating animal.

2. Any of various predatory, flesh-eating mammals of the order Carnivora, including the dogs, cats, bears, weasels, and raccoons.

Carrier Radio frequency capable of being modulated with some type of information.

Carrying Capacity 1. The steady-state density of a given species that a particular habitat can support. The number (or weight) of organisms of a given species and quality that can survive in, without causing deteriorating, a given ecosystem through the least favourable environmental conditions that occur within a stated interval of time.

2. The maximum population density of a given species in an area beyond which no significant increase can occur without damage occurring to the resources upon which the population depends.

3. The quantity of biomass an environment is capable of carrying over a 12-month period... or throughout the most critical period of the year ...or maintained indefinitely.

Carrying Capacity, Wildlife The user-specified-quality biomass of a population of a species or life group, under the influence of social or behavioural constraints, for which a particular area, having user-specified objectives, will supply all energetic and physiological requirements over a long, but specified, period. A distinction should be made for *actual* or *potential* carrying capacity. Actual carrying capacity varies annually. *Composite* carrying capacity implies many species (e.g., all of the grazing animals).

Catalogue number A five-digit number assigned to a catalogued orbiting object.

Catch Curve A graph representing the pattern of mortality of a population of fish over time resulting from fishing. Similar to a kill curve.

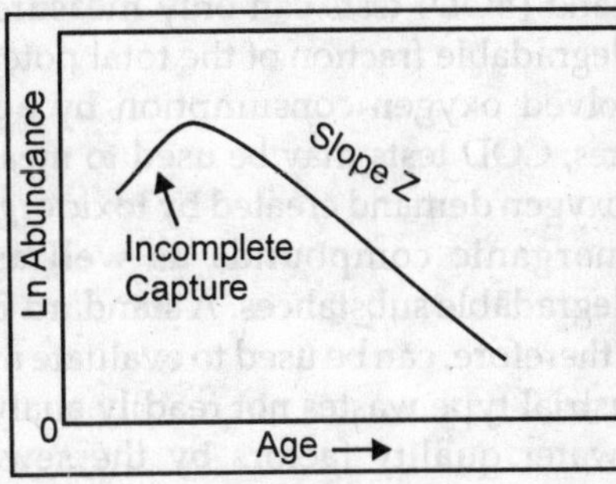

Fig. Catch Curve

Centigrade Temperature scale proposed by Swedish astronomer Anders Celsius in 1742. A mixture of ice and water is zero on the scale; boiling water is designated as 100 degrees. A degree is defined as one hundredth of the difference between the two reference points, resulting in the term centigrade (100th part). To convert centigrade to Fahrenheit; multiply the centigrade temperature by 1.8 and add 32oF = 9/5C + 32.

Central processing unit (CPU) Main part of a computer consisting of an arithmetic logic unit and a control unit.

Channel (Watercourse) An open conduit either natural or artificial, which periodically or continuously contains moving water, or which forms a connecting link between two bodies of water.

Channel Storage The volume of water that is temporarily present (stored) in a stream channel and its floodplain while enroute to an outlet.

Character Type Large physiographic area of land which has common characteristics of land forms, rock formations, water forms, and vegetative patterns.

Chemical Oxygen Demand (COD) A measure of the amount of a dissolved oxygen supply in a body of water that would be used up in completely oxidizing added inorganic oxidizable compounds, e.g., in the oxidation of ammonia to nitrate. Biological oxygen demand (BOD) tests can only measure the biodegradable fraction of the total potential dissolved oxygen consumption by added wastes; COD tests may be used to measure the oxygen demand created by toxic organic or inorganic compounds as well as by biodegradable substances. A standard COD test, therefore, can be used to evaluate many industrial type wastes not readily analyzed for water quality factors by the sewage-oriented BOD test.

Chemical Tissue Analysis Measurement of the types and/or amounts of chemical substances present in the tissue of an organism.

Chemosynthetic Activity The synthesis of organic matter from mineral substances, with the aid of chemical energy.

Chlorofluorocarbon (CFC) A family of compounds of chlorine, fluorine, and carbon, entirely of industrial origin. CFCs include refrigerants, propellants for spray cans and for blowing plastic-foam insulation, styrofoam packaging, and solvents for cleaning electronic circuit boards. The compounds' lifetimes vary over a wide range, exceeding 100 years in some cases.

CFCs' ability to destroy stratospheric ozone through catalytic cycles is contributing to the depletion of ozone worldwide. Because CFCs are such stable molecules, they do not react easily with other chemicals in the lower atmosphere. One of the few forces that can break up CFC molecules is ultraviolet radiation, however, the ozone layer protects the CFCs from ultraviolet radiation in the lower atmosphere. CFC molecules are then able to migrate intact into the stratosphere, where the molecules are bombarded by ultraviolet rays, causing the CFCs to break up and release their chlorine atoms. The released chlorine atoms participate in ozone destruction, with a single atom of chlorine able to destroy ozone molecules over and over again.

International attention to CFCs resulted in a meeting of diplomats from around the world in Montreal in 1987. They forged a treaty that called for drastic reductions in the production of CFCs. In 1990, diplomats met in London and voted to significantly strengthen the Montreal Protocol by calling for a complete elimination of CFCs by the year 2000.

Chlorophyll The green photosynthetic substance in plants which allows them to capture solar energy.

Choropleth Map A map showing discrete areas, such as counties, soil units, and vegetation types. Uniform conditions are assured within each map unit.

Circadian Rhythm The cyclical changes in physiological processes and functions that are related to the 24-hour *diurnal* cycle.

Circuit The complete path of an electric current; an assemblage of electronic elements; a means of two-way communication between two points—comprised of associated "go" and "return" channels.

Circularly polarized RF Radio frequency transmissions where the wave energy is divided equally between a vertically- and a horizontally-polarized components.

City Planning (Urban Planning) The process of guiding growth, maintenance, and change in urban areas to achieve a full range of objectives, well beyond the form and arrangement of the parts of the physical urban environment.

Clarke Belt A belt 22,245 miles (35,800 kilometers) directly above the equator where a satellite orbits the Earth at the same speed the Earth is rotating. Science fiction writer and scientist Arthur C. Clarke wrote about this belt in 1945, hence the name.

Classification Forming, sorting, apportioning, grouping, or dividing objects into classes to form an ordered arrangement of items having a defined range of characteristics, relationships, and distinctive differences. Classification systems may be taxonomic, mathematical, observed, or inferred, depending upon the purpose to be served. The purpose of classification is to describe the structure and relationships of objects to each other and to similar objects, and to simplify these relationships in such a way that general statements can be made about classes of objects, thereby achieving economy of memory and ease of manipulation.

Clay Typically naturally occurring inorganic, crystalline mineral particles less than 0.002 millimeters in equivalent diameter. Composed of fragments of hydrous aluminum silicate minerals in soil and other earthy deposits. "Soil texture" consisting of 40 percent or more clay-sized particles, less than 40 percent silt-sized particles and less than 45 percent sand-sized particles. Such soils have a high plasticity index, an expression of the amount of water that soil can hold without losing its plasticity.

Cleaning Cut Cleaning is applied to forest stands not past the sapling stage (4-8 feet), but not until the removal cut has been completed. At this stage, genetic differences are likely to be easily recognized; opportunity to improve species composition is at a maximum; and little production of wood on usable stems will be lost – with the possible exception of areas where rainfall exceeds vegetative needs.

Climate The average weather condition in an area determined over a period of years. The average course or condition of the weather at a particular place over a period of many years as exhibited in absolute extremes, means, ranges and seasonal distribution of air temperature, wind velocity and direction, precipitation type, duration and amount, humidity and other weather elements. It is not synonymous with weather.

Climatology Science dealing with *climate* and climate phenomena.

Climax Community (Climax) The final stable, self-perpetuating community in a developmental series under the prevailing climatic and soil conditions. A "climax" is determined in large measure by the nature of the topography or soil, e.g., a particular type of forest may be the "climax" on a north-facing slope while a grassland may be the "climax" on the south-facing slope of the same ridge. The main biotic components are not overthrown by invaders. No new species become dominant in the community. Climax is frequently used in the sense of species usually present (or at least common) only as members of a climax community.

Cline A gradual and continual change in a structural or functional character exhibited by a series of populations or throughout the range of a species, usually along a line of geographic or environmental gradient, in which individuals at the two extremes differ markedly.

Clinometer An instrument used to determine the height of a tree or structure.

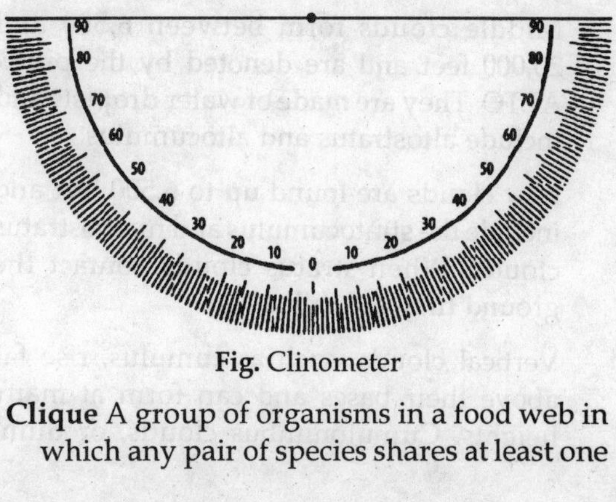

Fig. Clinometer

Clique A group of organisms in a food web in which any pair of species shares at least one prey species; species groups that have similar diets but are different from other such groups.

Clone A person or thing very much like another, e.g., a copy of another manufacturer's computer.

Cloud A visible mass of liquid droplets suspended in the atmosphere above Earth's surface. Clouds form in areas where air rises and cools. The condensing water vapour forms small droplets of water (0.012 mm) that, when combined with billions of other droplets, form clouds. Clouds can form along warm and cold fronts, where air flows up the side of the mountain and cools as it rises higher into the atmosphere, and when warm air blows over a colder surface, such as a cool body of water.

Clouds fall into two general categories: sheetlike or layer-looking stratus clouds (stratus means layer) and cumulus clouds (cumulus means piled up). These two cloud types are divided into four more groups that describe the clouds' altitude.

High clouds form above 20,000 feet in the cold region of the troposphere, and are denoted by the prefix CIRRO or CIRRUS. At this altitude water almost always freezes so clouds are composed of ice crystals. The clouds tend to be wispy are often transparent, and include cirrus, cirrocumulus, and cirrostratus.

Middle clouds form between 6,500 and 20,000 feet and are denoted by the prefix ALTO. They are made of water droplets and include altostratus and altocumulus.

Low clouds are found up to 6,500 feet and include the stratocumulus and nimbostratus clouds. When stratus clouds contact the ground they are called fog.

Vertical clouds, such as cumulus, rise far above their bases and can form at many heights. Cumulonimbus clouds, or thunderheads, can start near the ground and soar up to 75,000 feet.

Cloud streets Lines or rows of cumuliform clouds.

Cloudburst Any sudden, heavy rain shower.

Cluster Development Grouping structures in a housing development tightly together and using the open land thus saved for common greens and squares. Subdivision and zoning regulations apply to the project as a whole rather than to its individual lots (as in most tract housing). Densities are calculated on a project-wide basis. Potential advantages include: improved site design free of standard lot pattern limitations; lower street and utility costs made possible by reduced frontages; more useful open space due to reducing or eliminating the unusable side and front yards required by traditional zoning; greater flexibility in mixing residential building types; the possibility, through its greater freedom in design, of increasing over-all densities without loss of essential amenities.

Clutch The aggregate of eggs or the young of birds.

Coastal Waters Those waters, adjacent to the shorelines, which contain a measurable quantity or percentage of sea water, including, but not limited to, sounds, bays, lagoons, bayous, ponds, and estuaries.

Coastal Zone Colour Scanner (CZCS) The first spacecraft instrument devoted to measurement of ocean colour. Although instruments on other satellites have sensed ocean colour; their spectral bands, spatial resolution, and dynamic range were optimized for geographical or meteorological use. In the CZCS, every parameter is optimized for use over water to the exclusion of any other type of sensing. The CZCS flew on the Nimbus-7 spacecraft.

Coastal Zone The coastal waters (including the lands therein and thereunder) and the adjacent shores (including the waters therein and thereunder), strongly influenced by each other and in proximity to the shorelines of the several coastal states, including transitional and intertidal areas, salt marshes, wetlands, and beaches.

Co-dominant A tree receiving full sunlight from above but comparatively little from the sides. Such trees have medium-sized crowns.

Coefficient A constant factor of an equation or model as distinguished from a variable.

Cohort A group of plants or animals in the same age class. This is frequently discussed as the equivalent of 1,000 or 10,000 animals starting life together and being subject to different mortality factors, reducing the number over time.

Cold-Water Fishery Stream and lake waters that support predominantly cold-water species of game or food fishes, which have maximum, sustained water temperature tolerances of about 70 degrees Fahrenheit in the summer. Salmon, trout, grayling, and northern pike are examples.

Collector Road Roads and trails constructed to serves smaller land areas than a forest arterial road, and they are usually connected to a terminal or forest arterial or public highway. Collects traffic from forest local roads and/or terminal facilities. The location and standard are influenced by both long-term multi-resource service needs, as well as travel efficiency. May be operated by either constant or intermittent service, depending on land use and resource management objectives for the area.

Colluvium Mixed deposits of soil material and rock fragments accumulated near the base of steep slopes through gravity, i.e., soil creep, landslides, and local surface runoff (but not stream flow).

Colour (of Water) Colour in water can be caused by the presence of such things as plankton, decaying organic matter, industrial wastes, and sewage. "True colour" - the colour of a water sample after turbidity has been removed by filtration. "Apparent colour" - the colour of an untreated water sample. True colour is usually measured by comparing the colour of a water sample to that of a fixed standard. Colour is expressed in terms of "colour units" where one colour unit is the difference in tint produced by one milligram per litre of the chlorplatinate ion.

Comma cloud Band of organized cumuliform clouds that look like a comma from a satellite's perspective. Comma clouds are indicators of heavy storms.

Command An English-language specification issued by the user to a computer instructing it to perform some function.

Commensurability Capacity of a rangeland use permittee's base ranch property to support permitted livestock during the period that such livestock are off public land.

Commensurable Values Resource yields or conditions which can be objectively compared because their values can be expressed on the same measurement scale (e.g., in dollars, pounds, or cubic feet).

Commercial Forest Land Forest land that is producing or is capable of producing crops of industrial wood (20 cubic feet or more per acre per year) and 1. has not been withdrawn by Congress, the Secretary, or the Chief;.

2. existing technology and knowledge is available to ensure timber production without irreversible damage to soils, productivity, or watershed conditions; and.

3. existing technology and knowledge, as reflected in current research and experience, provides reasonable assurance that adequate restocking can be attained within 5 years after final harvesting (there is no financial criterion).

Commercial Thinning Thinning operations from which the material cut can be sold on the market as opposed to pre-commercial thinning in which small wood is left in the forest.

Commercial/Industrial Land Land used primarily for buying, selling, and processing goods and services and including sites for stores, factories, shopping centers, and industrial parks, together with necessary adjacent facilities such as underground and surface utilities, access streets and alleys, and other servicing structures, appurtenances, and measures.

Commodity A transportable resource product with commercial value; all resource products which are articles of commerce.

Common Variety Minerals Materials which lack substantial individual, energy-related, or metalliferous value and which are commonly extracted in significant quantities and which are often used in construction or related industries. Examples include but are not limited to sand, stone, gravel, and clay and combinations thereof. They cannot be claimed or patented.

Community (Ecological Community) A community is an assemblage of populations living in a stated area. It is a loosely organized unit to the extent that it has characteristics additional to its individual species and populations components. Communities have perceived functional unity, characteristic trophic structures, and patterns of energy flow. They also have taxonomic unity (in that there is a certain probability that certain species will occur together) and a relatively uniform appearance. Communities may be sharply defined. Very frequently, however, communities blend gradually into one another but they have the characteristic of being distinctively (statistically) different from adjacent units. The word is used with the simplest (e.g., an unrooted mat of algae) to the most complex ecosystem (e.g., a multistoried rain forest). Community units may be very large, like the continent-wide coniferous forest, or very small, like the community of invertebrates and fungi in a decaying log. The extent of a community is limited only by the requirement of a more or less uniform species composition. The composition and character of a community is an indicator of the type of environment that is present. Since "plant communities" and "animal communities" occur together in the same habitat and have many interrelations, the one can scarcely be considered independently of the other. Together they make up the "biotic community", a part of the general concept of community that also includes abiotic subsystems and factors.

Community Development Deliberate efforts to foster sociocultural change on the human community level, change of a kind seen as valuable or progressive, usually including an improved quality of life through increases in resources, skills, facilities, technology, social competence, and social power.

Community, Biotic An assemblage of related populations of plants and animals constituting a relatively self-sufficient ecological unit. Any assemblage of populations living in a prescribed area; it is a unit to the extent that it has characteristics additional to its individual species and populations components.

Compaction Decrease in volume of soil and sediment as a result of compressive stress, continued deposition above them, but also from drying, use of heavy equipment, and other causes.

Compartment A subdivision of forest area for purpose of orientation, administration, and management operations and usually defined by natural or artificial boundaries such as ridgelines, streams, or roads. Usually containing about 1,000 acres, a compartment is further divided into stands.

Compatible Uses Land uses which can exist together, so that no one use detracts significantly from the quality, output, or value of another.

Compensable Regulation Land use regulations which, while restricting the uses which may be made of land, compensate the property owner for any decrease in the value of his or her land caused by the regulations. The property is typically first valued as in a condemnation proceeding. Once a fair value is ascertained and agreed upon, the state guarantees the owner that value for his or her land. The owner then agrees to place the land under regulations controlling its use. This is a means of strengthening zoning by providing compensation for loss of control or value.

Competition (Ecological Competition) The general struggle for existence (and dominance) in which organisms compete for a limited supply of the necessities of life. It is the condition of rivalry between different organisms, usually closely related species (or within a species) which use the same resources and live in the same places, that exists when organisms are in short supply. If the resources are not in short supply, the condition that occurs when the organisms seeking and using that resource nevertheless harm one another in the process. Competition may be interspecific (i.e., between two or more different species), or intraspecific (i.e., between members of the same species).

Complement To supply mutually what another user or site lacks. (A design for a site complex that indicates the specific locations of buildings, parking lots, etc. complements the management plans that do not contain this information). The distinction between complements and supplements of plans is that complements, and the larger plan to which they relate, can be viewed as being mutually supplemental because each supplies something which the other lacks; however, plan supplements do not gain anything themselves by their addition to a plan.

Complementary Uses Land uses, each of which is improved in quality or quantity by the other uses existing with it in the same area.

Complete Flow Complete flow means that flow (stream discharge) figures are determined based upon a complete record (full range) of stage, as opposed to partial record flow where figures pertain only to a predetermined limited range in stage.

Complete Protection The withdrawal of all grazing animals from a given range.

Completeness Test Requirement that an alternative plan for water and related land resources provide and account for all necessary investments or other actions that will be needed to assure the full realization of the contributions provided by the plan to the components of the objectives specified for the planning area.

Component A functional part of an ecosystem. It can be a species, a part of a species, a life form, or any other functional group made up of other components (e.g., the component "grass" is constituted of many species in a given ecosystem, but it is defined by the ecosystem analyzer as a functional component based on an explicit rationale.).

Comprehensive Plan (General Plan, Master Plan) An official document adopted by a local government or landowner setting forth its general policies regarding the long-term physical development of a city or other area. Such plans are broad enough to include all aspects of a development or redevelopment program as distinguished from sporadic, isolated, or piecemeal planning. The process of designing futures not only of three dimensional accomplishments in space but also social mechanisms such as laws, regulations, policies, and forms of organization. It is used interchangeably with the terms "general plan" and "city plan" and is probably, most familiarly, as a "master plan". Comprehensive plans" typically include 1. a land plan map which shows the location of all owned land, the interrelationship between all of the uses proposed on such land, the circulation system required to serve the uses existing or proposed, and the relationship between ownership within a reasonable area of influence.

2. Land use policies related to the resources management, conservation and disposal of public lands.

3. Land use policies related to the activities and the regulations as they affect portions of the land.

4. Implementation schedules which establish priorities for programs in management, mapping, engineering, and public works.

5. A program for coordinating land use, intensity, and activity programs with central, state, regional, county and other local agencies as well as with private applicants. Increasing emphasis has occurred on social, economic, and esthetic topics, moving plans beyond physical and land use documents.

Computer Electronic machine capable of performing calculations and other manipulations of various types of data, under the control of a stored set of instructions. The machine itself is the hardware; the instructions are the program or software. Depending upon size computers are called mainframe, minicomputers, and microcomputers. Microcomputers include desk top and portable personal computers.

Concentration The amount of a substance occurring in a given amount of a medium — air, water, soil, tissue, etc. It may be expressed e.g., as parts per million, grams per litre, or in other units.

Condemnation In real property law, the process by which property of a private owner is taken for public use without his or her consent, but requiring payment of just compensation.

Condensation Change of a substance to a denser form, such as a gas to a liquid. The opposite of *evaporation*.

Conduction The transfer of heat from one substance to another by direct contact. Denser substances are better conductors; the transfer is always from warmer to colder substances.

Confined Aquifer An aquifer bounded above and below by impermeable beds or beds of distinctly lower permeability than that of the aquifer itself. It contains confined ground water.

Conifer Any of predominantly evergreen, cone-bearing trees with needle, awn, or scalelike leaves, e.g., pine, spruce, hemlock, cypress, larch, or fir (often called "softwoods").

Conservation The protection, improvement, and use of natural resources according to principles that will assure their highest economic and social service. A general word suggesting practices and customs of people that perpetuate resources and yields from some that are sustained. It usually includes preventing waste of non renewable resources and recycling. Said by some to be "wise use without abuse," definitions of "wise" and "abuse" remain to be defined in each situation of using such a phrase.

Constancy The percentage of occurrence of a species in the total number of plots, uniform in area, located in a number of forest stands of one kind of community.

Constrained Maximum Level Alternative The highest level of a particular output that could be produced over time, subject to the production of minimum acceptable levels for all other outputs.

Constraint A qualification of the minimum or maximum amount of an output or cost that could be produced or incurred in a given time period or with given factors or resources. A limiting or satisfying condition which must be met when setting objectives. "Ceiling" or "floor" may be used in the same manner.

Consulting Forester A self-employed professional forester.

Consumer In an ecosystem, a heterotrophic organism (one for which complex materials, especially organic foods, are the chief source of nutrition) which feeds on other organisms. A primary consumer, as a herbivore, obtains its nutrition directly from plants; a secondary consumer, as a carnivore, obtains its energy indirectly by feeding upon herbivores.

Consumer Scale Wood is hauled off the land and is measured at the mill (where the logger sells his/her wood). The logger reports this mill scale to the landowner. Payment to the landowner is based on this scale at the mill.

Consumptive Wildlife Species All wild creatures taken by humans for sport, food, or profit.

Contiguous tracts or any land units or legal subdivisions having a common boundary. Lands having only a common corner are not considered contiguous. *Contiguity* is the general condition and indices possibly expressing habitat or land use diversity.

Continent One of the large, continuous areas of the Earth into which the land surface is divided. The six geographically defined continents are politically defined as seven; Africa, Asia, Australia, Europe, North America, South America, and Antarctica.

Contour Stripping A type of strip mining that is practiced in areas of steep topography where the mineral seam outcrops or approaches the surface at approximately the same elevation along a hillside.

Contrails Condensation trails. Artificial clouds made by the exhaust of jet aircraft.

Convection The rising of warm air and the sinking of cool air. Heat mixes and moves air. When a layer of air receives enough heat from the Earth's surface, it expands and moves upward. Colder, heavier air flows under it which is then warmed, expands, and rises. The warm rising air cools as it reaches higher, cooler regions of the atmosphere and begins to sink. Convection causes local breezes, winds, and thunderstorms.

Cool Season Forage Plants that mainly provide a winter or early spring food source for wildlife (e.g., clover, winter wheat, rye, ryegrass).

Cooperage Wooden barrels tubs or the wood used in making them.

Coordinated Universal Time (UTC) (Greenwich Mean Time) Local time at zero degrees longitude at the Greenwich Observatory England. UTC uses a 24-hour clock, i.e., 2:00 a.m. is 0200 hours, 2:00 p.m. is 1400 hours, midnight is 2400 or 0000 hundred hours.

Coppice 1.Tree sprouts;.

2. A forest grown wholly or predominantly from sprouts from harvested trees.

Cordwood Small diameter wood suitable for pulp, chips, or firewood but not for sawlogs.

Core Drilling The process by which a cylindrical sample of rock and other strata is obtained by using a hollow-drilling bit that cuts and retains a section of the material penetrated.

Coriolis force The apparent tendency of a freely moving particle to swing to one side when its motion is referred to a set of axes that is itself rotating in space, such as Earth. The acceleration is perpendicular to the direction of the speed of the article relative to the Earth's surface and is directed to the right in the northern hemisphere. Winds are affected by rotation of the Earth so that instead of a wind blowing in the direction it starts, it turns to the right of that direction in the northern hemisphere; left in the southern hemisphere.

Corridor 1. A linear strip of land identified for present or future location of transportation or utility rights-of-way within its boundaries.

2. A thin strip of vegetation used by wildlife and potentially allowing movement of biotic factors between two areas.

Cost Efficiency The usefulness of specified inputs (cost) to produce specified outputs (benefits). In measuring cost efficiency, some outputs (such as environmental, economic, or social impact) are not assigned monetary values, but are achieved at specified levels in the least cost manner. Certain outputs not highly desired may be gotten at low costs. "Effective" emphasizes achievement of objectives. Cost efficiency is usually measured using present net value, although use of benefit-cost ratios and rates-of-return may be appropriate.

Coupled system Two or more processes that affect one another;.

Cover Plants and plant parts, living or dead, on the surface of the ground. Also the percentage of a named area that has such plants.

Cover (Forage Density) 1. The proportion of the ground occupied by a perpendicular projection to the ground from the outline of the aerial parts of plants.

2. Canopy closure or the amount of overhead tree cover as measured using a spherical densitometer held at 1.5 m from the ground, 2. The plants or other objects used by wild animals for nesting, rearing of young, resting, escape from predators, hiding from hunters, or protection from adverse environmental conditions.

3. Night time is the profound cover.

Cover Type Mappable classes of predominant vegetative cover or the map itself. Sometime used synomously with land use maps.

Cover-Forage Ratio The ratio of the area in a habitat analysis unit providing good cover to that area providing good forage.

Covert A distinctive dense vegetation in which game or other species of interest are typically found e.g., a corner in hedgerows.

Crook Sharp bend defect in a tree reducing its potential value.

Crop calendar The schedule of the maturing and harvesting of seasonal crops.

Crop Tree A well-located tree of potential high value which will be grown to maturity and will not be removed before the final harvest.

Cross Subsidy Selling a valuable timber stand for less than market value to compensate the purchaser for having to pay for another timber stand that is worthless.

Crown The branches and foliage of a tree; the upper portion.

Crown Class In descending order of crown height and size: dominant, co-dominant, intermediate, and suppressed.

Cruise 1.A survey of forest land to locate timber and estimate its quantity by species, products, size, quality, or other characteristics;.

2. The estimates from such a survey.

Crusing Radius (r) The radius of a mapped circular area in which average individuals typically move. Time, seasons, or totals may be specified. The circle (or concentric circles) may be drawn to encompass proportions of all observations of the animals. Where A = r^2, the radius is an expression of the area commonly used by individuals, families, or groups of animals of a species.

Cryosphere One of the interrelated components of the Earth's system, the cryosphere is frozen water in the form of snow, permanently frozen ground (permafrost), floating ice, and glaciers. Fluctuations in the volume of the cryosphere cause changes in ocean sea-level, which directly impact the atmosphere and biosphere.

Cubic Foot 1,728 cubic inches of wood; contains 6-10 useable board feet of wood.

Cull 1.A tree or log of merchantable size rendered unmerchantable because of poor form, limbiness, rot, or other defect;.

2. The deduction in gross volume made to adjust for defect;.

3. To cut a small portion of a stand by selecting one or a few of the best trees;.

4. To reject a tree, log, or board in scaling or grading;.

5. Any item of production that does not meett standards.

Culmination The point at which a satellite reaches its highest position or elevation in the sky relative to an observer (aka the closest point of approach).

Cycle Fairly regular fluctuations in density of a species over many years (>30). Cycles need to be evaluated in terms of the regularity of their *period* (the time between peaks or " lows") and also their amplitude (the difference from the average density).

Cyclone An area of low pressure where winds blow counterclockwise in the Northern Hemisphere and clockwise in the Southern Hemisphere.

D

Data Base A group of data stored together and organized according to characteristics that the data have in common.

Data collection system (DCS) DCS units are flown on both GOES and NOAA polar-orbiting spacecraft. They gather and relay data from both mobile and stationary platforms at various locations. DCS units on NOAA satellites can also determine the precise location of moving platforms at the time the data were acquired.

Data Family Those data records vertically related as one branch of an hierarchical tree.

Data rate The amount of information transmitted per unit time.

Datum Any level surface, line, or point used as a reference in measuring elevations. Also any single observation. "Data" is the plural.

Death Rate 1. Deaths within a cohort divided by the number of individuals in that cohort at the beginning of an age or time period;.

2. Proportions of deaths in a population.

Debark To remove the bark from trees or sections of trees, usually mechanical rather than manual.

Decibel (dB) A tenth of a bel. A unit used to measure the volume of a sound, equal to ten times the common logarithm of the ratio of the intensity of the sound to the intensity of an arbitrarily chosen standard sound. The decibel also is used to measure relative strengths of antenna and amplified signals and always refers to a ratio or difference between two values.

Deciduous Trees Thise which lose all of their leaves during the winter season.

Decision Variable A component of an alternative in which activities and their costs, outputs, and benefits are identified and used for analysis and decision making. All activities and costs necessary to accomplish the outputs and benefits are included.

Declination The angular distance from the equator to the satellite, measured as positive north and negative south.

Decomposer An organism, usually a bacterium or a fungus, that breaks down the bodies of dead plants and animals into simpler substances usable by green plants. These organisms mineralize organic matter.

Decreaser Plant Species (Decreaser) Plant species of the original vegetation that will decrease in relative amount with continued overuse, e.g., grazing.

Deep Chiseling Deep chiseling is a surface treatment that loosens compacted mine spoils. The process creates a series of parallel slots on the contour in the spoil surface that impedes water flow and markedly increases water infiltration.

Defect Ant irregularity or imperfection in a tree, log, piece product, or lumber that reduces the volume of sound wood or lowers its durability, strength, or utility value. Defects in lumber may result from insect or fungus attacks, growth conditions and abnormalities, manufacturing or seasoning practices, etc.

Defence Meteorological Satellite Program (DMSP) A U.S. Air Force-managed meteorological satellite program with satellites circling in sun-synchronous orbit. Imagery is collected in the visible- to near-infrared band (0.4 to 1.1 micrometers) and in the thermal-infrared band (about 8 to 13 micrometers) at a resolution of about three kilometres. DMSP data is available directly from the satellite for local use aboard ships and at military deployment locations, but is also usually available to civilian users.

Degradation Any significant reduction in the fertility of a soil. Also, general lowering of land surfaces by erosion.

Degree A unit of angular measure represented by the symbol o. The circumference of a circle contains 360 degrees. When applied to the roughly spherical shape of the Earth for geographic and cartographic purposes, degrees are each divided into 60 minutes.

Delphi Method A technique which allows people to arrive at a consensus about an issue of interest. It consists of a series of repeated interrogations, usually by means of questionnaires, of individuals whose opinions or judgments are of interest. After the initial interrogation of each individual, each subsequent interrogation is accompanied by information about the preceding round of replies. Each participant is thus encouraged to reconsider and, if appropriate, change his or her previous reply in light of the replies of other members of the group.

Delta The fan-shaped area at the mouth or lower end of a river; formed by eroded material that has been carried downstream and dropped in quantities larger than can be carried off by tides or currents.

Demand Estimates of the quantity of various outputs the public wants and would use relative to specific geographic areas. The amount of an output that users are willing to take at a specified price, time period, and condition of sale. The quantity of a service or commodity needed to meet the requirements of the user.

Demand Schedule The relationship between price and quantity demanded. The demand schedule expresses how much of the good or service would be bought or consumed at various prices at a particular point in time.

Demand Species Native and desired non-native species with high social, cultural, or economic values.

Demodulation The process of retrieving information (data) from a modulated carrier wave, the reverse of modulation.

Demography The study and description of human population structure and dynamics.

Den A rain proof, weather-tight cavity in a tree or a space dug by animals among rocks or in soil.

Den Tree A tree having one or more weather-tight cavities in its trunk or major limbs. The cavity serves as denning or escape cover for animals.

Dendrology Study of the identification, habits, and distribution of trees.

Density The actual number of individuals of a defined group occurring in a specified unit of space and time.

Density, Stand Number of individual plants of interest per unit of area, commonly expressed for forests as numbers of trees or stems per acre.

Density-Dependent Factors Any environmental factor that is dependent on populations density to be very effective. Example: Floods kill all animals in an area (assumed 100%). It is density independent. Some diseases kill many or few animals, dependent on the contagion and stresses related to the number of animals present.

Depletion Changes in the time distribution of physical rates of use (usually use exceeds replenishment). The redistribution of use is in the direction of the present, i.e., more use (or emphasis on availability of use) of a resource is to be made for times closer to the present than on use later.

Derived Demand The demand (schedule) for a commodity that grows out of the desire to satisfy the demand for some other commodity. The demand for housing, for example, may create a demand for lumber, bricks, and many other things needed to build houses.

Descending node The point in a satellite's orbit at which it crosses the equatorial plane from north to south.

Desert A land area so dry that little or no plant or animal life can survive.

Desertification The man-made or natural formation of desert from usable land.

Design The description (verbal, graphic, model, etc.) which shapes, limits, and controls the physical development and changes that will occur in time and space. Design sets the general guidelines for the character of development. Individual project level design sets the precise character, pattern and form of the project environment.

Desirable Plant Species Rangeland plant species of moderate to high palatability that are preferred by animals or beneficial in soil and water conservation.

Desired Condition A statement describing a common vision for a specific area of land or type of land within the plan area. Desired conditions are aspirations and are not commitments or final decisions. Statements of desired conditions include the estimated time required for their achievement. They may be achievable only over a long time period. The social, economic, and ecological attributes toward which management of the land and resources of the plan area is to be directed. They also take into account the range of natural variability typical for the landscape, the uncertainty of natural disturbances, the effects of past management, the unique features or opportunities that forests and grasslands can contribute, and the human desires and possible uses of the land.

Desired Non-native Species Those species of plants or animals that are not indigenous to an area but which represent a significant, and usually remnant segment of a gene pool.

Detector A device in a radiometer that senses the presence and intensity of radiation. The incoming radiation is usually modified by filters or other optical components that restrict the radiation to a specific spectral band. The information can either be transmitted immediately or recorded for transmittal at a later time.

Detention Storage Surface runoff which is temporarily present in ground surface irregularities (or even on level surfaces) while it evaporates or flows overland toward stream channels.

Detergents A term applied to a wide variety of cleansing agents used to wash clothes, dishes, and other articles. Generally, organic materials that are surfactants in aqueous solutions.

Detritivore (Detrivore) An organism (plant or animal) that feeds on organic remains or organic (mostly plant) debris.

Developed Recreation Recreation that requires facilities that result in concentrated use of an area. Examples of development are areas such as campgrounds and ski areas. Facilities might include: roads, parking lots, picnic tables, toilets, drinking water, ski lifts, and buildings.

Developer Any person or persons who directly or indirectly, through any formal or informal combination or aggregation, own or control a tract or tracts of land for which such person or persons propose a project. "Person" includes any individual, partnership, corporation, association, unincorporated organization, trust, estate, or any other legal or commercial entity, except central, state or local government agencies. Specifying the developer as financially responsive makes certain that individuals cannot establish dummy corporations, trusts, or otherwise organize in a manner to avoid the size and time period limitations of the definition of projects (such as using an organization to hide two separate tracts comprising more than 25, but less than 50, units; selling or developing two such tracts which are functionally interrelated but separated by a short distance; selling or developing one such tract, waiting a few years, and developing a second tract).

Development A built-up or intensive-use tract of land, especially used in the sense of a subdivision having the necessary utilities — i.e., water, gas, electricity, paved areas, roads, etc. that utilizes the land only as a location, not for its natural resource properties. There may be types such as residential, commercial, or industrial.

Development Rights A broad range of less-than-fee-simple ownership interests. It includes mainly various forms of the easement, a clearly defined property right. The landowner keeps the title but agrees to continue using the land as in the past, the right to develop the land resting in the holder of the easement.

Development Type Those land use types by which the formerly-open spaces of an area are prominently altered or made to appear built-up, i.e., buildings, paved areas, and other types of site alterations.

Dew Atmospheric moisture that condenses after a warm day and appears during the night on cool surfaces as small drops. The cool surfaces cause the water vapour in the air to cool to the point where the water vapour condenses.

Dew point The temperature to which air must be cooled for saturation to occur; exclusive of air pressure or moisture content change. At that temperature dew begins to form, and water vapour condenses into liquid.

Diagrammatic Representation (Blob Diagram, Schematic Diagram) An analytical technique in which the physical components (such as parking lots, garbage cans, picnic tables, beach etc.) in a plan are represented symbolically, usually just by their names written inside individual circles. The relationships between components then represented by lines connect those that have necessary or desirable support functions. The individual components are then shifted to minimize the extent to which different interconnecting lines cross each other. The resulting solution is assumed to represent the most efficient spatial relationship between project components, and thus represent idealized goal of the actual physical plan. Often, utilized at the project or site planning level, such diagrams may be used for studying the optimal relationships between vaguely defined components of a land use plan.

Diameter at Breast Height (DBH) The diameter of a tree measured four feet, six inches from the ground level. "Ground level" can follow a convention; either the highest point off the ground touching the stem, or the mean of the highest and low points.

Diameter Tape A tape calibrated to measure the diameter of a tree by measuring its circumference.

DIB Diameter inside bark, usually measured at the small end of a log.

Digital image An analogue image converted to numerical form so that it can be stored and used in a computer. The image is divided into a matrix of small regions called picture elements or pixels. At sub-satellite point each pixel represents a specific amount of area. For example, in APT each pixel represents 4.1 kilometres. Each pixel has a numerical value or data number value, quantifying the radiance of the image at that spot. The data number value of each pixel usually represents a value between black and white, i.e., shades of grey.

False colour can be applied to the image by assigning a graduated colour palette to represent the grey shades. The colour is "false" because it represents an assigned, not actual, colour.

Digital system A system in which information is transmitted in a series of pulses. The source is periodically sampled, analyzed, and converted or coded into numerical values and transmitted. Digital transmissions typically use the binary coding used by computers so most data is in appropriate form, but verbal and visual communication must be converted. Many satellite transmissions use digital formats because noise will not interfere with the quality of the end product, producing clear and higher-resolution imagery.

Dimension Lumber Hardwood dimension lumber processed so that it can be used virtually in the sizes provided in the manufacture of furniture or other products. Softwood dimension lumber is boards more than 2 inches thick but less than 5 inches thick. Such wood is used in construction and is sold by units such as 2x4s, 4x8s, or 2x10s.

Direct Effect (Primary Effect) A condition caused by an action or inaction without an intermediary causal agent; an effect characterized by a close causal relationship.

Direct readout The capability to acquire data directly from environmental satellites via an Earth station. Data can be acquired from NOAA and other nations' environmental satellites, which offer weather information from geostationary and polar-orbiting satellites.

Disadvantaged Groups Lacking in the basic resources or conditions (as standard housing, medical and educational facilities, civil rights) believed to be necessary for an equal position in society. Once "poor" and "underprivileged", the phrase covers more than the economic conditions associated with "poor" and does not have the implication of social or economic oppression contained in "underprivileged.".

Discharge The volume of water flowing past a point per unit time, commonly expressed as cubic feet per second, million gallons per day, gallons per minute, or cubic meters per second.

Discharge, Thermal Commonly a discharge of heat energy from an industrial facility, usually in the form of heated air or water as a byproduct of operations.

Discing Land preparation for crops or reforestation by pulling a harrow with large discs over a site; a type of earth scarification and competing plant removal.

Discount Rate The interest rate used in formulating plans and evaluating the results of discounting future benefits and computing costs, or otherwise converting benefits and costs to a common time basis. It represents the cost or time value of money in determining present value of future costs and benefits.

Discounted Benefits Total dollar benefits produced over a given period of time discounted to the present and expressed in current dollars.

Discounted Costs Total dollar costs accrued over a given period of time discounted to the present and expressed in current dollars.

Discounting An adjustment, using a discount rate for the values of money over time so that costs and benefits occurring in the future are reduced to a common point in time, usually the present, for comparison. The practice of placing a lesser value (economic or other) on future events than on present events for the purpose of comparison. An item received today is seen to be worth more than an identical item received next year. Discounting refers only to the timing of an event and should not be confused with reduced values based on the uncertainty of future events or on implied quality changes over time.

Dispersion 1.The pattern or distribution of organisms in space,.

2. The movement of organisms in space from a point,.

3. The movement of organisms in time,.

4. A condition following such acts.

Dissolved Oxygen (DO) DO in water is expressed in milligrams per litre and the amount present reflects chemical, physical, and biological activities in the water body. It can only be increased by aeration and the photosynthetic processes of aquatic plants.

Dissolved Oxygen Concentration (Dissolved Oxygen Level, DOL) The amount of oxygen dissolved in a water sample expressed in parts per million. In unpolluted water, dissolved oxygen is usually present in amounts of 10 parts per million or less. Adequate dissolved oxygen is necessary for the life of fish and most other aquatic organisms. About 3 to 5 parts per million is the lower limit needed for the support of fish life over a long period of time.

Dissolved Solids The total amount of solid matter (mainly inorganic salts) that remains after a filtered volume of water has been evaporated. The amount of residue of organic and inorganic matter is commonly expressed as a proportion of the original sample in parts per million parts of the original.

Distance Zones Areas of landscapes denoted by specified distances from the observer. Used as a frame of reference in which to describe landscape characteristics or activities of people and described as foreground (fg), middle ground (mg), or background (bg).

Disturbance Processes Actions, functions, or events that influence or maintain the structure, composition, or function of the terrestrial or aquatic components of ecosystems. Natural disturbances include, among others, drought, floods, -wind, fires, insects, and pathogens. Human-caused disturbances include actions such as recreational use, livestock grazing, mining, road construction, timber harvest, land-use development, and the introduction of exotic species.

Disturbed Land Terrestrial surface which has been altered —either physically, biologically, or chemically— by the action of people, e.g., land on which excavation has occurred or upon which overburden has been deposited, or both.

Diurnal Performed in twenty-four hours, such as the diurnal revolution of the Earth.

Diurnal arc The apparent arc described by heavenly bodies from their rising to their setting.

Diversion Taking significant quantities of water from a stream or other body of water into a canal, pipe, or other conduit. Applies to groundwater stations when pumping is significant.

Diversion Ditch A channel with a supporting ridge on the lower sside constructed across a slope to intercept surface water.

Diversity 1. The distribution and abundance of different plant and animal communities and species within the area covered by a land and resource management plan.

2. The relative degree of abundance of wildlife species, plant species, communities, habitats, or habitat features per unit of area.

3. The total range in kinds and numbers of species potentially related in a defined area.

4. A numerical index which incorporates the total number of species in a biotic community and some rating of the relative importance (numbers, biomass, productivity, etc.) of individual species. Two commonly used indices are the Simpson and Shannon-Weiner indexes. See Biodiversity.

Dobson Unit (DU) The standard way to express ozone amounts in the atmosphere. One DU is 2.7 x 10^{16} ozone molecules per square centimetre. One Dobson unit refers to a layer of ozone that would be 0.001 cm thick under conditions of standard temperature (0°C) and pressure (the average pressure at the surface of the Earth). For example, 300 Dobson units of ozone brought down to the surface of the Earth at 0°C would occupy a layer only 0.3 cm thick in a column. Dobson was a researcher at Oxford University who, in the 1920s, built the first instrument (no called the Dobson meter) to measure total ozone from the ground.

Doldrums Region near the equator characterized by low pressure and light shifting winds. See wind.

Dominance, Ecological The condition in communities or in vegetation strata in which one or more species, by means of their numbers, coverage, or size, have considerable influence or control upon the conditions of existence of associated species.

Dominance, Social The determination of the behaviour of one or ore animals by the aggressive behaviour or otherwise of other individuals, resulting in establishing a social hierarchy.

Dominant The land against which an easement or privilege exists is called the servient tenement and the estate to which it is annexed is the *dominant* tenement and their owners are called respectively servient and dominant owner. A dominant plant typically receives the greatest amount of sunlight in a community. It is usually the most visually conspicuous plant but other characteristics or traits may make "greatest influence" the meaning.

Dominant Use A concept in which land and water resources are classified and managed for one main type of use. Other uses may or may not be excluded but are always of secondary importance to the optimum development for the dominant use.

Dominant Use Management Management based on the idea that although land may be capable of many uses, it will provide for one use better than any other. The land is managed for the single purpose of maximizing that use to the exclusion of other uses where conflicts exist. Devoting each land unit to its best use is assumed to be the most efficient allocation of resources. This assumption has been shown to be erroneous by "operations research", procedures, among others.

Doppler effect (Doppler shift) The apparent change in frequency of sound or light waves, varying with the relative velocity of the source and the observer. If the source and observer draw closer together; the frequency is increased. Named for Christian Doppler; Austrian mathematician and physicist (1803-1853).

Doppler radar The weather radar system that uses the Doppler shift of radio waves to detect air motion that can result in tornadoes and precipitation, as previously-developed weather radar systems do. It can also measure the speed and direction of rain and ice, as well as detect the formation of tornadoes sooner than older radars.

Dormant Seedling Seedlings that have ceased visible growth due to high or low temperature, moisture, or other causes.

Downconverter Any radio frequency circuit that converts a higher frequency to a lower frequency This enables signal processing by a receiver. A typical downconverter will feature one or more states of RE preamplification, a mixer where the frequency conversion occurs, a local oscillator chain, and often one or more intermediate frequency preamplifiers to minimize the effect of line losses between the converter and the receiver.

Drag A retarding force caused by the Earth's atmosphere. Thus by definition, drag will act opposite to the vehicle's instantaneous velocity vector with respect to the atmosphere. The magnitude of the drag force is directly proportional to the product of the vehicle's cross-sectional area, its drag coefficient, its velocity, and the atmospheric density and inversely proportional to its mass. The effect of drag is to cause the orbit to decay or spiral downward. A satellite of very high mass and very low cross-sectional area, and in a very high orbit, may be very little affected by drag, whereas a large satellite of low mass, in a low altitude orbit may be affected very strongly by drag. Drag is the predominant force affecting satellite lifetime.

Dragline An excavating machine that utilizes a bucket operated and suspended by means of lines or cables, one of which hoists or lowers the bucket from a boom; the other cable, from which the name is derived, allows the bucket to swing out from the machine or to be dragged toward the machine for loading.

Drain A small artificial watercourse designed to drain swampy areas or irrigated lands. Theoretically, it is actually a small canal, but it is referred to as a "drain" in many localities.

Drainage Area The drainage area of a stream at the specific location of the site is that area, measured in a horizontal plane, enclosed by a topographic divide from which direct surface runoff from precipitation would normally drain by gravity into the stream above the site.

Drainage Basin The area from which water is carried off by a drainage system, a watershed, or a catchment area.

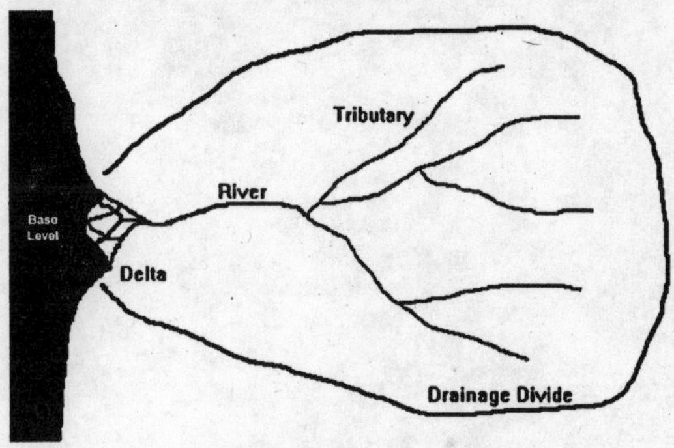

Fig. Drainage Basin

Duff Forest litter and other organic debris in various stages of decomposition on top of the mineral soil, typical of conifer forests in cool climates where rate of decomposition is slow and litter accumulation excedes decay.

Durability (Environmental Durability) The relative ability of resources to tolerate sustained use, without degradation or destruction of the resource base (i.e., lowering productivity or desirable resource quality). This ability is sometimes rated on a qualitative scale extending from "fragile" to "durable". Durability in land use planning refers to those resource types or the evaluation process considers properties that are resistant to degradation in the presence of whatever types. For example, vegetated dunes may be classed as "environmentally fragile" for off-road vehicle uses while unvegetated dunes are "environmentally durable".

Dynamic Programming A multistage decision process that systematically searches out a sequence of decisions which maximizes or minimizes some predefined objective function. The method is based on Bellman's Principle of Optimality which states that "an optimal policy has the property that whatever the initial state and initial decision are, the remaining decisions must constitute an optimal policy with regard to the state resulting from the first decision.".

Dynamics The study of the action of forces on bodies and the changes in motion they produce.

Dynamics, Population Description of the change in abundance and structure of plnt or animal populations resulting from births, deaths, migrations, and related rate phenomena.

Dystrophic Pertaining to lakes with brown water, high humic material and organic matter content, low nutrient availability, poor bottom fauna, and high oxygen demand; oxygen is continually depleted and pH is low. In lake aging, the "age" between a eutrophic lake and a swamp.

E

Earth Observing System (EOS) A series of small- to intermediate-sized spacecraft that is the centerpiece of NASAs Mission to Planet Earth (MTPE). Planned for launch beginning in 1998, each of the EDS spacecraft will carry a suite of instruments designed to study global climate change. MTPE will use space-, aircraft-, and ground-based measurements to study our environment as an integrated system. Designing and implementing the MTPE is, of necessity, an international effort. The MTPE program involves the cooperation of the U.S., the European Space Agency (ESA), and the Japanese National Space Development Agency (NASDA). The MTPE program is part of the U.S. interagency effort, the Global Change Research Program.

Earth Observing System Data & Information System (EOSDIS) The system that will manage a dataset of Earth science observations to be collected over a 15-year period. Existing data indicates that the Earth is changing, and that human activity increasingly contributes to this change. To monitor these changes, a baseline of "normal" performance characteristics must be obtained. For the Earth, these baseline characteristics must cover a global scale and a long enough period that the variation caused by seasonal changes and other cyclical or periodic events (e.g., El Nino and the solar cycle) may be included in the analyses. The baseline characteristics also must enable scientists to quantify processes that govern the Earth's system. Functionally EOSDIS will provide computing and networking facilities supporting EOS research activities, including data interpretation and modeling; processing, distribution, and archiving of EDS data; and command and control of EGS observatories.

Earth Probes Discipline-specific satellites and instruments that will be used by NASA to obtain observations before the launch of EDS spacecraft. Generally smaller than the EGS satellites and instruments, Earth Probes are planned to complement the broad environmental measurements from EOS with highly focused studies in areas such as tropical rainfall (TRMM), ocean productivity (SeaWi ES), atmospheric ozone (TOMS), and ocean surface winds (NSCAT).

Earth system science An integrated approach to the study of the Earth that stresses investigations of the interactions among the Earth's components in order to explain Earth dynamics, evolution, and global change.

Easement An interest or legal right in the land of another person that allows the easement holder to engage in specified uses or rights without actually owning the land. The person or estate against which the easement or privilege exists is called the servient tenement and the estate to which it is annexed is the dominant tenement and their owners are called respectively *servient* and *dominant* owner.

Easement, Conservation A group of negative and affirmative easements that keep land permanently in open use or a natural state, or provide some other conservation or recreation function. These include scenic, flooding, and agricultural easements. Easements may include the following provisions: 1. Prohibitions against erecting buildings or other structures;.

2. Restrictions against constructing or altering any private drives or roads;.

3. Prohibitions against removing or destroying trees, shrubs, or other greenery;.

4. Restrictions against uses other than residential or agricultural, for public utilities, and existing uses;.

5. Restrictions against displaying outdoor signs, billboards, or any other form of outdoor advertising;.

6. Prohibitions against dumping or storing trash, wastes or unsightly or offensive materials of any kind;.

7. Other kinds of restrictions consistent with open space preservation and reservation.

Easement, Continuing An easement which is self-perpetuating, independent of human interventions, as the flow of a stream, or one which may be enjoyed without any act on the part of a person.

Easement, Discontinuous An easement, the enjoyment of which may be had only by the interference of a person, e.g., a right-of-way or a right to draw water.

Easement, Negative An interest in land which can give the easement-owner the right to prevent the owner of the land from doing something on it, e.g., scenic easements.

Easement, Recreational Affirmative easements allowing public entry on the servient land to hunt, fish, or enjoy other recreational activities.

Easement, Scenic Essentially an easement of view. Scenic easements restrict the use of land adjacent to public highways, parks, and rivers. They may provide something attractive to look at within the easement area, an open area to look through to see something attractive beyond the easement itself, or a screen to block out an unsightly view beyond the easement area.

Eclipse The partial or total apparent darkening of the sun when the moon comes between the sun and the Earth (solar eclipse), or the darkening of the moon when the full moon is in the Earth's shadow (lunar eclipse).

Eclipse blindness Focus-point type of vision loss caused by looking at the sun for too long a time, which can burn a hole in the retina of the eye.

Ecoclass A classification system for the biological and earth sciences based on linking together existing disciplinary classifications of the major ecosystem components. It is a unifying framework for linking three basic biological and earth science stratifications or classifications: the vegetation subsystem, the land subsystem, and the aquatic subsystem. A hierarchical structure is given in which the lowest level units are the "ecological land unit" (ELU), a composite of the land and vegetation subsystems, and the "ecological water unit" (EWU), a composite of elements of the land and aquatic subsystems.

Ecological Composition The biological components of an ecological system, which are the foundation of diversity at the genetic, species, and landscape scales. Genetic diversity is the variation in inheritable characteristics within and among individual organisms and populations. Species diversity is the number and different kinds of species present in a given area. Landscape diversity is the variety of plant communities (including their identity, distribution, zuxtaposition, and seral stage) and habitats evaluated at the landscape scale.

Ecological Conditions Components of the biological and physical environment that can affect diversity of plant and animal communities and the productive capacity of ecological systems. These components could include the abundance and distribution of aquatic and terrestrial habitats, roads and other structural developments, human uses, and invasive, exotic species.

Ecological Determinism A particular point of view in planning which advocates that retaining natural elements and processes should be given priority in planning. Natural elements (such as aquifers, wildlife breeding areas, and flood plains) which are valued by segments of society or that freely perform necessary life support or quality of life functions, or whose alteration will significantly damage other valued resources, should be protected in plans. Ecological determinism is a synonym for physiographic determinism, biological primacy, and biophysical determinism. Although "ecological" strictly implies concern with organisms and their environmental relationships, "ecological determinism" also applies to retaining such nonliving environmental elements as aquifers, steep slopes, and severe storm impact zones.

Ecological Land Type (ELT) An area of land with a distinct combination of natural, physical, chemical, and biological properties that cause it to respond in a predictable and relatively uniform manner to the application of given management practices. In a relatively undisturbed state and/or at a given stage (sere) of plant succession, a predictable and relatively uniform plant community usually occupies an ELT. Typical size generally is in the mid-hundreds of acres.

Ecological Land Type Phase These are subdivisions of those ELT's where vegetation management is most common. They share the same characteristics as ELT's; however, their size is smaller (10-100 acres) and the biological and physical conditions are more limited. They are also known as Forest Habitat Types.

Ecological Land Unit (ELU) One of the lowest levels of the "Ecoclass" system of classifying subdivisions for forest management. Units of land having strong slope steepness; aspect; rock types and conditions, geomorphology, soil characteristics and productive capabilities, type, density and age of vegetation, and drainage character. The basic physical unit of land that scientific disciplines agree must be delineated and examined as a separate entity (for evaluating or management purposes), and analysis of onsite potentials, capabilities and limitations. It is the most significant level of land stratification which best communicates the basic (inherent) capabilities and limitations. These are land (or water) units which, because of their strong uniformity in physical and biological characteristics, respond similarly to management activities or other stimuli. They are sometimes called response units.

Ecological Management Unit (EMU) A segment of the landscape which contains soils with similar characteristics, enabling it to be managed as a single body. A coded map unit of the Soil Resource Inventory. Each EMU is based on five components - landform, source of materials, texture, water regime, and modifiers, and reflects soil capabilities. The primary component is landform.

Ecological Niche (Much debated) The role a particular organism plays in the environment.

Ecological Pyramids Groups of organisms in nature often consist of many small organisms associated with a few larger organisms. This distribution of number and size has been compared to a pyramid with a broad base representative of numerous small organisms and the point of the pyramid representing a few largest organisms.

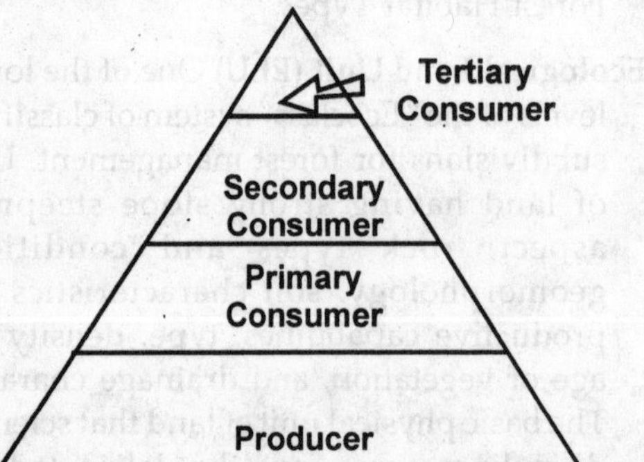

Fig. Ecological Pyramids

Ecological Stability (Balance of Nature) An ecological system may be said to be stable during that period of time when no species becomes extinct (thereby creating a vacant niche) and none reaches plague proportions for long enough to destroy the niches of other species and cause them to be come extinct. A stable ecosystem is one in which all niches are fully occupied by appropriate species. Whatever stability there is in the ecological world it is not a static equilibrium, but a fluctuating or dynamic one, one normally a highly fluctuating. Stability lies in the ability to bounce back, not in the ability to hold tenaciously to ground once taken or numbers once achieved.

Ecological Succession The orderly, likely sequential transition of an area through distinctively different stages of development.

Ecological Sustainability The maintenance or restoration of ecological system composition, structure, and function which are characteristic of a plan area over time and space, including but not limited to ecologicaL processes, biological diversity, and the productive capacity of ecological systems.

Ecological Water Unit (EWU) The lowest levels of the "Ecoclass" system of classifying divisions for forest description and management. An EWU is a composite of elements from the land and aquatic subsystems, where aquatic type and the adjacent land types together define a homogeneous unit.

Ecology Science dealing with the inter-relationships between living organisms and their environments. 1. A branch of science dealing with the relationships of organisms and their cycles and rhythms, community development, and environments — especially as manifested by natural structure, interaction between different kinds of organisms, geographic distributions, and population alterations.

2. The study of the relation of organisms or groups of organisms to their environment.

3. The study of the relations among living organisms and their environment.

4. The scientific discipline that is concerned with the relationships between organisms and their past, present, and future environments. These relationships include physiological responses of individuals, structure and dynamics of populations, interactions among species, organization of biological communities, and processing of energy and matter in ecosystems.

Econometrics Testing and quantifying of economic theories and the solution of economic problems by their conversion to mathematical forms, applying mathematical processes, and using statistical techniques.

Economic Analysis Studying the economic effects of a given action or decision. The effectiveness of various proposed resource management alternatives or combinations of alternatives are studied using the estimation of costs, output volumes, and values for the outputs etc. and how they may also vary each year in an expected pattern of flow in the future. It includes measurement of all pertinent desirable and undesirable consequences to all segments of the economy stated in comparable, monetary terms. Monetary values are discounted to a common time. It may also include an analysis of factors not quantifiable in monetary terms.

Economic Determinism A point of view which advocates that planning decisions should be based primarily upon the results of analysis of changes in resource supply and demand in response to changes in the "price" for the right to use a resource. That potential user or use which is willing or able to give up the most wins the use rights. Maximizing the total economic welfare is considered to result in maximizing the benefits to society from resource use allocation and this is viewed a being the goal of all resource use. Total economic welfare will be maximized by providing the resources to the highest bidders in such quantities that the price individuals are willing to pay exactly equals the cost of providing the last unit of the resource, under standard economic assumptions. The view that economic factors are fundamental or dominant, despite the importance of biological, psychological, political, cultural, and social factors.

Economic Enterprise (Economic Unit) A business of sufficient resources to provide an acceptable standard of living for a family, often termed an economic unit.

Economic Externalities Effects of resource allocation decisions which are not included in the market mechanism, and as such have no economic influence on resource allocation decisions; these constitute failures of classical market mechanism as a resource allocation tool. Also, costs or benefits that occur whenever the activities of one or more persons affect the welfare or production functions of others who have no direct control over that activity.

Economic Growth Increased financial activity in real terms over time. It may include building infrastructure, resources, and public relations. It does not necessarily correspond to increased production of physical goods.

Economic Life The time extending from the date of installation [of a facility] into service to that date when the facility is no longer economically profitable to use.

Economic Scarcity (Resource Scarcity) The condition in which a limited number of marketable goods are desired in higher quantities than are available. The "scarcity" can be reduced either by increasing the available goods and maintaining a low price, or by raising the price of the goods until the number of people willing to pay the higher price are at the supply level. Which is the socially superior method depends on supply costs and the existence of economic externalities.

Ecosystem (Bioc(o)enosis, Biogeoc(o)enosis, Biogeoc(o)enose) Any natural unit or entity including living and non-living parts that interact to produce a stable system through cyclic exchange of materials. 1. A complete, interacting system or unit of organisms in a space considered together with their environment, e.g., a marsh, a watershed, a lake, etc., mentally isolated for purposes of study or management. A flow of energy leads to clearly defined food and feeding relationships, biologic diversity, and biogeochemical cycles (i.e., exchange of materials between living and nonliving parts) operating as an integrated system. "Ecosystem" is the preferred term in English while "biocoenosis" or "biogeocoenosis" is preferred by writers using or familiar with the Germanic and Slavic languages.

2. A nameable natural unit that includes not only the total array of plant and animal species in an environment, but also the energy to power the system.

3. A biotic community plus its abiotic environment.

4. Any complex of living organisms with its environment.

5. An interconnected community of plants and animals, including humans, and the physical environment within which they interact.

Ecosystem Diversity The variety and relative extent of ecosystem types, including their composition, structure, and processes within all or a part of an area of analysis.

Ecosystem Dynamics Characteristic and measurable processes and associated change within an ecosystem such as succession, energy flow, nutrient cycling, and community metabolism.

Ecosystem Integrity The completeness of an ecosystem that, at multiple geographic and temporal scales, maintains its characteristic diversity of biological and physical components, spatial patterns, structure, and functional processes within its approximate range ofhistoric variability. These processes include disturbance regimes, nutrient cycling, hydrologic functions, vegetation succession, and species adaptation and evolution. Ecosystems with integrity are resilient and capable of self-renewal in the

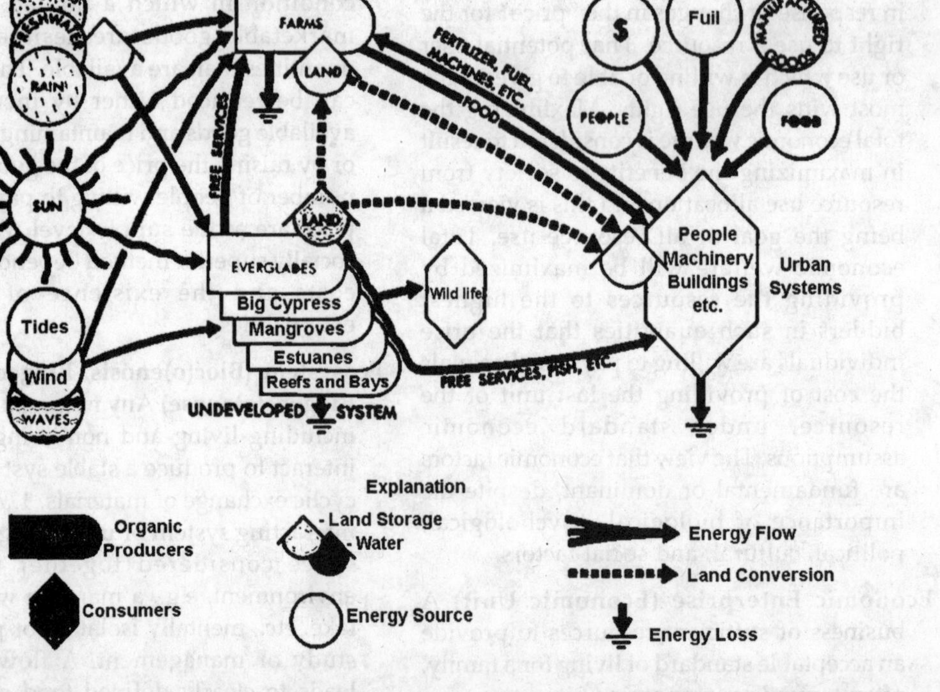

Fig. Ecosystem

presence of the cumulative effects of human and natural disturbances.

Ecosystem Stability 1. A measure of the tendency of an écosystem to persist, relatively unchanged, through time;.

2. A measure of the resilience of an ecosystem to perturbation.

Ecosystem Structure The biological and physical attributes that shape ecological systems; biotic attributes include population size, structure and range; fdliage~density and layering, snags, large woody debris or the size, shape and spatial relationships of cover types within a landscape; physical attributes include soil and geologic substrate variables, slope and aspect, or stream gradient.

Ecotone A transition between two communities or vegetation types. It is a junction zone narrower than the adjoining community areas themselves. The ecotonal community commonly contains many of the organisms of each of the overlapping communities and, in addition, organisms that are characteristic of and often restricted to the ecotone. Often, both the number of species and the population density of some of the species are greater in the ecotone than in the communities flanking it. Organisms which occur primarily or most abundantly or spend the greatest amount of time in contiguous communities are often called "edge species," while the tendency for increased species richness and density at community junctions is known as the "edge effect.".

Ecotourism Diverse tourism that include a major part of interest, activities, and successes related to ecological topics, environmental awareness and education, and experiencing the outdoor nature of places to which people travel. It includes the services and structures that support these activities. It includes the impacts, positive and negative, of such activities on the environment and human communities. Hector Ceballos-Lascurain, Mexico, is said to have coined the word *ecotourism* in 1983.

Ecotype A genetically distinct race or subdivision of species adapted to local habitat and climate. These significantly different genetic groups are broader than a biotype and narrower than a species. Ecotypic variation must be considered before transferring information on impacts to a species in one region to the same species in another region because different ecotypes may respond differently to the same level of effect.

Edaphic An adjective suggesting the soil complex or soil as a system.

Edaphic Factor (Soil Factor) A characteristic or factor of the soil environment (opposed to physiographic or climatic factors), which limits the existence or development of an organism or community, e.g., low nutrient or micro-nutrient concentrations, or the presence of toxic concentrations of certain substances, or insufficient or excess soil moisture conditions.

Edge The linear elements in a view or scene which are noted (or considered by the observer) as being the dividing lines between major components. "Edges" serve as visual boundaries e.g., land-water boundaries or forest-meadow boundaries. Also, for animal analyses, the volume that is the living space having length, width, height, and species-specific quality (or relevance to life requirements.).

Edge Effect The effect created in animal populations in the edge volume between two types. This portion of the forest environment, for example, provides a diversity of wildlife food and cover, where grass, forbs, and shrubs form a narrow border around forest clearings. Greater animal richness and abundance is the "effect".

Effect (Impact), Economic The change, positive or negative, in economic conditions, including the distribution and stability of employment and income in affected local, regional, and national economies, which directly or indirectly result from an actively, project, or program.

Effect (Impact), Physical, Biological The change, positive or negative, in the physical or biological conditions which directly or indirectly results from an activity, project, or program.

Effect (Impact), Social The change, positive or negative, in social and cultural conditions which directly or indirectly result from an activity, project, or program.

Effective PC Timber sale purchaser road costs that can be credited against the amount bid on a timber sale.

Effectiveness Index The index used in cost-effectiveness analysis, one emphasizing achieving objectives, not simply efficiency of any activity.

Effectiveness Test In choosing among alternatives in planning for water and related land resources, the "effectiveness test" refers to technical performance of the plan and the level of contribution to achieving each of the objectives.

Efficiency An expression of output per unit of input or cost.

Efficiency Test Using the *efficiency test* in choosing among alternatives in planning for water and related land resources requires that among all acceptable alternatives (central and non- central, water and non water, structural and nonstructural) a given alternative plan should be the least costly means (considering all adverse effects) of achieving the objectives.

Effluent Any liquid flowing out of the ground or from an enclosure to the surface flow network, typically the liquids discharged from domestic, industrial and municipal waste collection systems or treatment facilities.

Egalitarianism A belief or social philosophy that all people are equal in intrinsic worth and are entitled to equal access to the rights and privileges of their society. It may include leveling of social, political and economic inequalities. Exactly what aspects of social reality are considered in order to decide what should be equal and what unequal is a rather complex problem sometimes involves self contradictions: e.g., equality of opportunity can result in unequal distributions, because people have different abilities, degrees of motivation, and luck: equality of resource inputs in different schools may result in inequality in educational achievements of the students, (if the latter began with unequal advantages), and attempting to increase student achievement, different schools might require unequal resource inputs in order to overcome the initial handicaps of some students.

El Nino A warming of the surface waters of the eastem equatorial Pacific that occurs at irregular intervals of 2-7 years, usually lasting 1-2 years. Along the west coast of South America, southerly winds promote the upwelling of cold, nutrient-rich water that sustains large fish populations, that sustain abundant sea birds, whose droppings support the fertilizer industry Near the end of each calendar year; a warm current of nutrient-poor tropical water replaces the cold, nutrient-rich surface water Because this condition often occurs around Christmas, it was named El Niho (Spanish for boy child, referring to the Christ child). In most years the warming lasts only a few weeks or a month, after which the weather patterns return to normal and fishing improves. However; when El Nino conditions last for many months, more extensive ocean warming occurs and economic results can be disastrous. El Nino has been linked to wetter; colder winters in the United States; drier; hotter summers in South America and Europe; and drought in Africa.

Elasticity In economics, a measure of responsiveness of the quantity demanded or supplied to changes in price. Elasticity measures the degree to which price is effective in calling or holding back quantity. Small price changes do not influence demand for highly elastic commodities.

Electromagnetic radiation Energy propagated as time-varying electric and magnetic fields. These two fields are inextricably linked as a single entity since time-varying electric fields produce time-varying magnetic fields and vice versa. Light and radar are examples of electromagnetic radiation differing only in their wavelengths (or frequency). Electric and magnetic fields propagate through space at the speed of light.

Electromagnetic spectrum The entire range of radiant energies or wave frequencies from the longest to the shortest wavelengths – the categorization of solar radiation. Satellite sensors collect this energy, but what the detectors capture is only a small protion of the entire elctromagnetic spectrum. The spectrum usually is divided into seven sections: radio, microwave, infrared, visible, ultra-violet, x-ray, and gamma-ray radiation.

Electromagnetic wave Method of travel for radiant energy (all energy is both particles and waves), so called because radiant energy has both magnetic and electrical properties. Electromagnetic waves are produced when electric charges change their motion. Whether the frequency is high or low all electromagnetic waves travel at 300,000,000 meters per second.

Electromotive Producing an electric current through differences in potential. electromotive force The force that can alter the motion of electricity, measured in terms of the energy per unit charge imparted to electricity passing through the source of this force. Electromotive force causes current flow in a circuit.

Element set (Keplerian elements, classical elements, satellite elements) Specific information used to define and locate a particular satellite. The set includes the catalog number; epoch year; day and fraction of day; period decay rate; argument of perigee; inclination; eccentricity; right ascension of the ascending node; mean anomaly; mean motion; revolution number at epoch; and element set number. This data is contained in the two-line orbital elements provided by NASA in the NASA Prediction Bulletin.

Elevation The angle at which an antenna must be pointed above the horizon for optimal reception from a spacecraft.

Elites The individuals at the top stratification hierarchy, such as economic, social prestige, or politic. "Power elite" are those with a high degree of coordination and a set of common interests among those in top corporate and governmental positions.

Elliptical orbits Bodies in space orbit in elliptical rather than circular orbits because of factors such as gravity and drag. The point where the orbiting satellite is closest to Earth is the perigee, sometimes called per-apsis or perifocus. The point where the satellite is farthest from Earth is called apogee apoapsis, or api-focus. A line drawn from perigee to apogee is the line-of-apsides, sometimes called the major-axis of the ellipse. It's simply a line drawn through the ellipse the long way.

Eltonian Pyramid See Pyramid of Numbers.

Emergence (of a plant) 1. Appearance of the plant above ground after germination in the soil.

2. Elongation of the hypocotyl out of the soil into the air followed by the beginning of growth of the epicotyl.

Eminent Domain The right or power of government, subject to constitutional and statutory limitations, to take private property for public use after making just compensation.

Endangered Species Species listed as nationally in danger of extinction throughout all or a significant portion of their ranges. An animal that may become extinct because of loss or damage of habitat within the original range of the species.

Endemic Native or confined to a certain region; having a comparatively restricted distribution (as a disease caused by an indigenous pathogen).

Endemic Organism A taxonomic category (e.g., genus, species, and variety) the natural occurrence of which is confined to a certain region and the distribution of which is relatively limited.

Energetics The analysis and use of knowledge of embodied energy and the flow of energy in systems; the study and use of the knowledge of comprehensive energy budgets in designing and managing systems.

Energy budget A quantitative description of the energy exchange for a physical or ecological system. The budget includes terms for radiation, conduction, convection, latent heat, and for sources and sinks of energy.

Engineering Solution Development or assembly and use of technical solutions to solve (or at least minimize) problems associated with particular land uses or land use practices.

Enhancement Improvement on existing or projected natural biological conditions.

ENSO (El Nino-Southern Oscillation) Interacting parts of a single global system of climate fluctuations. ENSO is the most prominent known source of interannual variability in weather and climate around the world, though not all areas are affected. The Southern Oscillation (SO) is a global-scale seesaw in atmospheric pressure between Indonesia /North Australia, and the southeast Pacific. In major warm events El Nino warming extends over much of the tropical Pacific and becomes clearly linked to the SO pattern. Many of the countries most affected by ENSO events are developing countries with economies that are largely dependent upon their agricultural and fishery sectors as a major source of food supply employment, and foreign exchange. New capabilities to predict the onset of ENSO events can have a global impact. While ENSO is a natural part of the Earth's climate, whether its intensity or frequency may change as a result of global warming is an important concern.

Enteric Bacteria Those which originate in the intestines of warm-blooded animals.

Entomology, Forest Studies and practices dealing with insects in relations to forests and forest products.

Entrainment 1. The passage of organisms through the cooling systems of, for example, power plants.

2. Ambient water being brought into the cooling plumes as the effluents are discharged from power plants.

Environment The complex of physical, chemical, and biological factors in which a living organism or community exists.

Environment, Human The aggregate and interaction of all the external conditions and influences affecting the life and development of an individual or society.

Environmental Analysis Determining the probable effect of a project upon the renewable resources, considering the physical, biological, natural, social, and economic factors and their relations. It is also the process of preparing an environmental assessment or environmental impact statement and the decision about whether to prepare an environmental assessment or impact statement. It is an analysis of alternative actions and their predictable short-term and long-term environmental effects.

Environmental Assessment (EA) A public document that serves to 1. Provide sufficient evidence and analysis for determining whether to prepare an environmental impact statement or to find no significant impact and;.

2. Aid in agency's compliance with the rules when no environmental impact statement is necessary.

Environmental Corridor (Buffer Zone) A strip of land designed to protect the natural environment of an area and to prevent undesirable encroachment such as commercial development along a stream. Also a thin habitat connecting two similar areas and said to reduce "fragmentation.".

Environmental Effect An outcome, result, or consequence to the environment brought about by some force, project, or action. Primary (direct) impacts are generally caused by action inputs (such as road construction). Secondary (indirect) impacts commonly result from action outputs, (such as improved access leading to overload of recreation visitors).

Environmental Impact Analysis (Environmental Analysis) An activity that considers, estimates, and evaluates, using a systematic, interdisciplinary approach, the physical, natural, and social factors, the probable effects of a plan or proposed project upon local systems.

Environmental Modification Deliberate or unintentional changes produced by people that act upon an organism or community to influence its development or existence.

Environmental Planning Planning which considers the environment as a physical and biological system, one that is spatial, having direct experiential and sensory surroundings, and one that is a primary conditioner of physical and psychological development, health, and longevity for each human. It is strongly concerned with human activities taking place in the outdoors and those that reconcile their demands for and impacts upon the environment. It seeks to ensure that policies, programs, designs and actual developments incorporate forecasts of the effects upon environmental amenity and natural system functions and structures so as to anticipate, evaluate, and prepare for the consequences of decisions in light of present and future values.

Environmental Psychology That area of psychology which deals with environmentally induced behaviour or mental states of individuals or groups, e.g., where "behavioural psychology" may attempt to alter self concepts through operant conditioning, "environmental psychology" seeks to determine how the every-day physical environment affects human behaviour. It is also a study of human perception of elements in the environment or of different types of environmental settings. Perception includes direct sensory knowledge and cognitive discrimination expressed in terms of preference rated on a scale of desirability. It also includes learning and the knowledge of how a set of behaviours (e.g., littering, vandalism, and hunting) is changed.

Environmental Quality The perceived condition, the sum of the forces and factors which influence people's satisfaction and health in their work, leisure, living conditions, and community. Management, conservation, preservation, restoration, or improvement can enhance it in cultural and natural systems. Major systems of concern are areas of natural beauty; water, land, and air; biological resources and select ecosystems; geological, archeological, and historical; and resources perceived to be of great future value. "Quality" may be high or low. It may be present without people, but human perception, valuation, and judgement are needed to arrive at its expression. Individuals perceive it; a group expression (e.g., various statistical or consensus values) about it may be derived.

Environmental Science A multi- and inter-disciplinary science attempting to measure and evaluate the influence of people on the structure, function and relations of systems upon which humans depend, and the influence of management on these systems and their contributions to human objectives.

Environmental System A general phrase meaning a specified or defined set of interacting components in the environment, understood to have some degree of integrity, often having the characteristics of a general system. It may be a farm, a forest, or a watershed and is usually more general and inclusive of human factors than "ecosystem.".

EPA (Environmental Protection Agency) A U.S. agency that ensures: Federal environmental laws are implemented and enforced effectively; U.S. policy—both foreign and domestic—fosters the integration of economic development and environmental protection so that economic growth can be sustained over the long term; public and private decisions affecting energy, transportation, agriculture, industry, international trade, and natural resources fully integrate considerations of environmental quality; national effort to reduce environmental risk are based on the best available scientific information communicated clearly to the public; everyone in our society recognizes the value of preventing pollution before it is created; people have the information and incentives they need to make environmentally-responsible choices in their daily lives; and schools and community institutions promote environmental stewardship as a national ethic.

Ephemeral Stream (Intermittent Stream, Perennial Stream) A stream or portion of a stream that flows only in direct response to precipitation. It receives little or no water from springs and no long-continued supply (less than a month and often only during storm events) from snow or other sources. Its channel is at all times above the water table.

Ephemeris A tabulation of a series of points that define the position and motion of a satellite.

Epicormic Branching Branches which grow out of the main stem of a tree, arising from buds under the bark. Severe branching increases knottiness, thereby reducing the quality of lumber.

Epilimnion That region of a body of water that extends from the surface to the thermocline and does not have a permanent temperature stratification; the volume above the hypolimnion.

Epixylon Bacterial and fungal films on rocks and dead material in streams.

Epoch (Epoch Time or TO) Epoch specifies the time of a particular description of a satellite orbit.

Equator An imaginary circle around the Earth that is everywhere equally distant (900) from the North Pole and the South Pole. The equator is a great circle and defines latitude 00.

Equilibrium Price The price at which consumers will be willing to take exactly the amount that sellers want to place on the market at the price where marginal revenue of a product equals marginal cost.

Equilibrium, Ecological A dynamic state of quasi-balance in which the elements of a physical or biological system partition energy flow in a uniform manner, with outflow balancing inflow about which the system fluctuates to some small degree. (Often applied to an animal population at zero growth, to the steady interaction of two species, predators and prey, to the energy flow through an ecosystem, and to the nutrient cycling pattern of an ecosystem). (brief zero-rate-of-change in a fluctuating system).

Erodibility The relative ease with which one soil sample erodes under specified conditions as compared with other soils under the same conditions.

Erosion The group of processes whereby earth or rock material is worn away, loosened, or dissolved and removed from any part of the

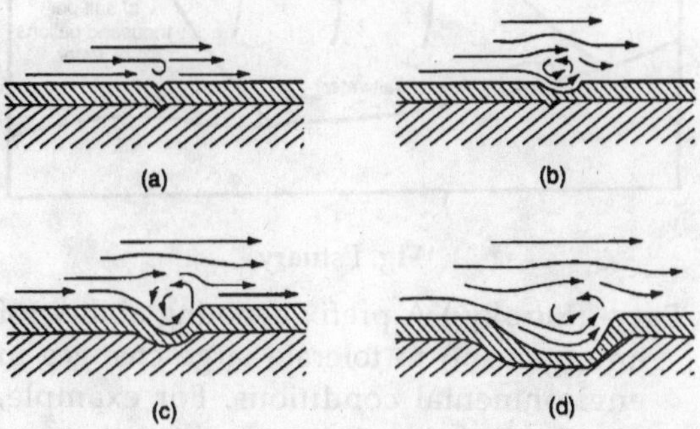

Fig. Erosion

Earth's surface by wind, water, ice, or gravity. It includes the processes of weathering, solution, corrosion, and transportation. It is often classified by the eroding agent (wind, water, wave, or raindrop erosion) and/or by the appearance of the erosion (sheet, rill, or gully erosion) and/or by the location of the erosion activity (surface, or shoreline) or by the material being eroded (soil erosion or beach erosion). "Raindrop erosion" always takes the form of "sheet erosion" though wind action or the movement of thin sheets of water can also cause "sheet erosion" over the ground surface. "Sheet", "gully" and "rill erosion" are all forms of "soil erosion." "Sheet" and "rill erosion" are the two forms of surface erosion. "Beach erosion" is always "shoreline erosion", e.g., because not all shorelines are beaches; "shore line erosion" is not always "beach erosion". The term accelerated erosion is used in comparing erosion caused by human activities with that occurring at natural rates (i.e., geologic erosion).

Erosion Hazard The possibility of a soil eroding when disturbed, based on soil type, topography, and vegetative cover.

Erosion, Accelerated Erosion that is much more rapid than natural erosion, or geologic erosion, and occurring primarily as a result of the influence of the activities of people or, in some cases, of animals or natural catastrophes that expose bare surfaces (for example, fires).

Erosion, Bank Destruction of land area by active cutting of stream-banks.

Erosion, Beach The retreat of the shoreline of large lakes and coastal waters caused by wave action, shore currents, or causes other than subsidence.

Erosion, Geologic (Natural Erosion) Gradual wearing away of the land surface which occurred prior to the occupancy of a land area by people or the present day wearing-away which is not due to the activities of people but due to natural environmental conditions of climate, vegetation, etc.

Erosion, Gully Removal of soil and formation of relatively large channels or gullies cut into the soil by concentrated surface runoff. Gullies are too deep to be oblitreated by ordinary tillage practices.

Erosion, Normal The gradual erosion of land used by people which does not greatly exceed natural erosion.

Erosion, Rill Removal of soil by the cutting of numerous small (several inches deep), but conspicuous, water channels or tiny rivulets by concentrated surface runoff usually from cultivated or exposed soils. It is intermediate between sheet erosion and gully erosion.

Erosion, Sheet The removal of a fairly uniform layer of soil from the land surface by runoff water or wind without the development of conspicuous water channels.

Erosion, Splash (Raindrop Erosion) The spattering of small soil particles caused by the impact of raindrops on wet soils. The loosened and spattered particles may or may not be subsequently removed by runoff.

ESA European Space Agency.

Escape Cover Vegetation dense enough to aid animals in escaping from potential enemies, including hunters; a synonym for security, security area, security cover, and hiding cover, but imprecise.

Escapement 1. The number, or proportion, of wildlife surviving the hunting season (not escape cover). The emphasis may be on specific age and sex classes.

2. Also fish passing a dam or waterway blockage.

Essential Elements A chemical element required by green plants for normal growth, such as the primary essential elements: hydrogen, oxygen, nitrogen, phosphorus, and potassium; secondary essential elements: sulphur, calcium, and magnesium; and the trace (or minor) elements: iron, boron, manganese, copper, zinc, and molybdenum. The last six, and traces of other elements, are required in only minute quantities.

Esthetics An acceptable spelling variation of aesthetics.

Estuarine Sanctuary A research area which may include any part or all of an estuary, adjoining transitional areas, and adjacent uplands, constituting to the extent feasible a natural unit, set aside to provide scientists and students the opportunity to examine over a period of time the ecological relationships within the area.

Estuary (Estuarine) All or part of the wide mouth of a navigable or interstate river or stream or other body of water having unimpaired natural connection with open sea tides and within which the sea water is measurably diluted with fresh water from land runoff.

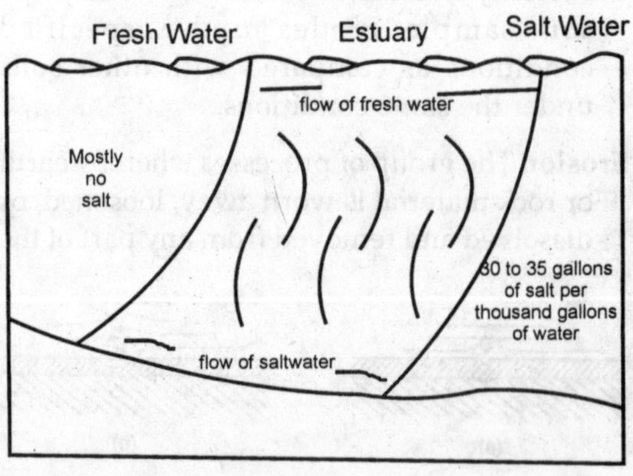

Fig. Estuary

Eury- (Euroky-) A prefix denoting ability (of an organism) to tolerate large changes in environmental conditions. For example, eurythermal describes organisms that can tolerate large temperature ranges.

Eutrophic Characteristic of waters rich in dissolved inorganic or mineral nutrient materials and exhibiting a paucity or complete lack of oxygen in the bottom waters as a consequence of high primary production and the high nutrient content.

Eutrophication The process whereby a body of water becomes rich in dissolved nutrients through natural or man-made processes. This often results in a deficiency of dissolved oxygen, producing an environment that favours plant over animal life.

Evaluating and sorting trees or logs according to quality.

Evaluation 1.Examination or judgement about the worth, quality, significance, amount, degree, or condition of something. Assessing or estimating value.

2. A comprehensive analysis of social, economic, and ecological conditions and trends relevant to a unit. The analysis of monitoring data that produces information needed to answer specific monitoring questions. Evaluation may include comparing monitoring results with a predetermined guideline or expected norm that may lead to recommendations for changes in management, a land management plan, or monitoring plan. Evaluations provide an updated compilation of information for use in environmental analysis of future project and activity decisions.

Evaporation Change from a liquid (more dense) to a vapour or gas (less dense) form. When water is heated it becomes a vapour that increases humidity Evaporation is the opposite of condensation.

Evapotranspiration Loss of water from a land area through transpiration of plants and evaporation from the plant and other surfaces.

Even-Aged Management The application of a combination of actions, a system, of growing, harvesting, and reproducing timber in stands of trees that grow together and, within each, are of essentially the same age (at least within 10-year classes). Managed even-aged forests are characterized by a distribution of stands, among which,there are varying ages (and, therefore, tree sizes throughout the forest area). Cutting methods producing even-aged stands are clearcut, shelterwood, or seed tree. 1. Clearcutting - The removal in a single cut of the entire standing crop of trees in a stand. It prepares the area for rapid seed germination and growth of a new even-aged stand.

2. Shelterwood Cutting - A method designed to establish a new crop under a remaining portion of the old stand that provides both a seed source and protection of the site and seedlings.

3. Seed Tree Cutting - The removal in one cut of the mature timber from an area with a small number of seed-bearing trees retained for regeneration. The difference in age between trees forming the main canopy level of a stand usually does not exceed 20 percent of the age of the stand at harvest rotation age. Regeneration in a particular stand is obtained during a short period at or after the time that a stand has reached the desired age or size for regeneration and is harvested.

Even-aged Regeneration Harvest A timber cutting procedure by which a new age class of trees is created; the methods include clearcutting, seed tree, shelterwood, and coppice.

Even-Aged Stand A forest stand composed of trees having no, or relatively small, differences in age. By convention the maximum difference admissible is generally 10 to 20 years, though with rotations of less than 100 years differences up to 25 percent of the rotation period may be admissible in some policy formulations.

Even-Flow Continuous supply of products over a given time period.

Evergreeen Trees which retain green foilage throughout the year. Not all conifers are evergrees (e.g., tamarack).

Exempt Stock Livestock that are grazed on public land free of charge. Usually confined to animals used for domestic purposes; saddle horses, milk cows, etc.

Existence Value (Bequest Motive) The external benefit of natural environments (or any kind of good but often of wilderness) that accrues to individuals having no intention of ever visiting the site or using the good in question. These people are willing to give up resources simply to know that the area, feature, or good exists in a particular condition. The concept of "existence value" has been recognized for some time. John Krutilla (1967) has used the term "bequest motive" to express roughly the same idea. Existence value is preferred because it is less specific in its connotations about motivations; i.e., it does not imply handing something down to later generations.

Existing Visual Conditions The present state of visual alteration which is measured in six degrees (untouched, unnoticed, minor disturbance, disturbed, major disturbance, drastic disturbance) of deviation from the natural appearing landscape.

Exosphere The uppermost layer of the atmosphere, its lower boundary is estimated at 500 km to 1000 km above the Earth's surface. It is only from the exosphere that atmospheric gases can, to any appreciable extent, escape into outer space.

Exotic Organism Any organism (plant or animal) that is not native in the area where it now occurs (introduced). "Invasive" may include "weedy" natives as well as invasive exotics. " A rose in a cornfield is a weed." Exotics may (or not) be invasives as can "native" plants.

Expectations The rules governing social situations, which are specific to particular societies and even to groups within societies. These are notable only when violated. Expectations are normative, in that they reflect what people believe ought to happen, and statistical, in that they express what people believe probably will happen. These may be probability statements or 1.0 minus the probability of a failure or bad outcome.

Experimental Forest A designated area managed for research purposes, e.g., for watershed or timber growth study.

Exploitation Utilization of a resource for expected profits or benefits; the number or proportion of animals removed by hunting, fishing or other means.

Export Base Analysis (Economic Base Analysis) A multiplier analysis method for estimating regional economic effects of projects. It is particularly suited to rural areas. The economy is divided into two sectors, "export" and "service". The export sector includes all goods and services produced for consumption outside the region being evaluated. The service sector includes all other economic activity. The export sector is assumed to create the need for the service sector, and a multiplier is calculated for each export industry, relating employment in the service sector to the level of activity in that industry. By determining project effects on the export industries, the total economic effects can be estimated.

Exposure (Slope Orientation, Aspect) The compass direction toward which the slope of a land surface faces. Transformations are recommended, i.e., 360o is likely to be equivalent to 1° or 2° or 358° or 359°.

Extant Extant plants and animals which are relics of the flora and fauna or other geologic time periods and are of historic or scientific significance are considered under botanical areas or zoological areas, respectively, and are identified and formally classified primarily because of their recreational values.

External forcing Influence on the Earth system (or one of its components) by an external agent such as solar radiation or the impact of extraterrestrial bodies such as meteorites.

Extinct Being no longer found anywhere in the world (e.g., passenger pigeon) without reference to cause(s).

Extirpation Eradication of a species (total) from an area.

Extractive Economics The economics of industries which remove natural resources and supply them to other industries as raw materials. The economic analysis of these industries is more often complicated by depletion of fixed stocks, resource ownership ambiguities, and "common pool" externalities than is true with other types of industries.

Extrinsic Resource Recreational values created by human-made changes, adaptations, and additions to basic natural (i.e., intrinsic) resources made by people, e.g., mill ponds, historic sites, tourist service facilities, archeological sites, etc.

F

Face Cord Stove length wood with a face of 32 square feet (a stack 4 ft high and 8 ft wide); the volume depends upon its length.

Facilities Capital improvements (typically long-lasting) such as buildings, roads, trails, fences, bridges, dams, heliports, and airfields that primarily facilitate producing resource elements.

Facsimile (FAX) A process by which graphic or photographic information is transmitted or recorded by electronic means.

Factor, Abiotic Environmental A non-living component of the environment (e.g., climate, slope, aspect) which may affect the living components of that environment.

Factor, Biotic Environmental Plants or animals that affect ecological processes, e.g., shading by trees or trampling by animals.

Factor, Controlling Environmental One or more limiting factors or conditions influencing an organism or assemblage of organisms in a way that accounts for a large part of the observed behaviour and distribution of that biological entity, e.g., the wilting of a plant because of insufficient soil moisture.

Factor, Density-Dependent An influence (occurrence or magnitude) that is responsive to the numbers of organisms per unit area (or volume) and as a result of such responses either enhances or reduces the suitability of the environment for the organism in question, e.g., a limited number of prey animals for the number of predators present in an area.

Factor, Density-Independent An influence that is effective without regard to the density of individuals in a population, e.g., very unfavourable weather.

Factor, Limiting Environmental The environmental influence by which the limit of toleration of a plant or animal population is first reached and which therefore acts as the immediate restriction to one or more of its functions or activities or in its geographic distribution.

Fahrenheit Temperature scale designed by the German scientist Gabriel Fahrenheit in 1709, based upon water freezing at 32°F and water boilmg at 212°F under standard atmospheric pressure. Compare with centigrade.

Family Unit A developed recreation site, including facilities and space needed for picnicking or camping, with a capacity of up to five people and generally suited for use by one average size family.

Far infrared Electromagnetic radiation, longer than the thermal infrared, with wavelengths between about 25 and 1000 micrometers.

Farmstead or Headquarters Land used primarily for dwellings, barns, pens, corrals, gardens, and other uses in connection with operating farms or ranches.

Fault (Earthquake Fault) A fracture in the ground surface or fracture zone along which there has been displacement of one side with respect to the other.

Fault, Active A linear break in the Earth's surface that has undergone movement in recent geologic time (the last 10,000 years) and may be subject to future movement.

Fauna The animal life of an area, at a time, used in the broad sense to include birds, mammals, fish, reptiles, amphibians, insects, mollusks, and crustaceans. Life, other than plants.

Feasibility (Use Feasibility) Evaluation of lands-use conditions based on the present likelihood or under specific management conditions with socioeconomic assumptions or climate (e.g., accessibility or population potential) or social climate. Areas close to a city have a higher feasibility than those more distant.

Fecundity Production of young by animals within an age class (typically a rate). Precisely, the conception rate; also the ability to produce offspring.

Feed Any non-injurious edible material having nutrient value; may be harvested forage, range or artificial pasture forage, grain, or other processed food for livestock or wild fauna.

Feedback A component of general systems theory. Negative feedback is a return to a system that reduces the quantity or quality of outputs. Positive feedback tends to increase subsequent outputs.

Feedback Loop The pathway within a system by which a portion of the output of a system or process returns to become a part of its inputs, changes the process, feed-forward, or system context. The adaptive or corrective steps in a management system.

Feedhorn A metallic cylinder closed at one end, used to obtain and direct radio frequency (RF) energy reflected from a satellite dish. It acts as a wave guide at microwave frequencies. RE energy inside the horn is picked up by a small probe; once inside the horn, the wavelength (energy) of the microwave radiation changes to a guided wave.

Felling (Cutting) Cutting down standing trees. "Felling" is not a synonym for logging or harvesting. These latter terms also include removing felled trees and some initial processing of the timber prior to removal.

Feng Yun Chinese geostationary environmental satellite that was destroyed by an explosion before launch in April 1994. The name Feng Yun, meaning Wind and Cloud, was originally applied to the Chinese polar-orbiting environmental satellite launched in September 1991 (Feng Yun 1-2), which offered direct readout services. The Chinese polar-orbiter program has since been abandoned.

Fertility Production of male or female sex cells or average male or female viable sex cell production.

Field The set of influences (electricity magnetism, gravity) that extend throughout space.

Field of view The range of angles that are scanned or sensed by a system or instrument, measured in degrees of arc.

Filter Device that while selectively passing desired frequencies removes undesired ones.

Filter Strip A strip of land of variable width retained or constructed along live streams as a safeguard to prevent sediment reaching the stream.

Final Cutting The logging operation that removes the last of the old crop trees in a stand that is being managed under the shelterwood system of silviculture.

Fire Break A natural or artificial barrier usually created by removing vegetation to prevent or retard the spread of fire. (Also called a fire line or fire lane.).

Fire Management All activities required for protecting resources and values from fire, and using fire to meet land management goals and objectives.

Fire Management Area One or more parcels of land with clearly defined boundaries and with established fire management direction that is responsive to land and resource management goals and objectives.

Fire Planning Planning for predicting, preventing, detecting, and influencing the spread and control of wildfires.

Fish All species of fresh or salt-water fishes, as well as crustaceans, mollusks, and other underwater organisms that are considered part of the fishery resource. Exclusive of marine mammals. Inclusion (or not) of sea turtles is now decided in each case.

Fish, Game (sport fish) Those species of fish considered possessing sporting qualities on fishing tackle. Examples of freshwater game fish are salmon, trout, grayling, black bass, muskellunge, walleye, northern pike, and lake trout.

Fish, Rough A non-sport fish, usually omnivorous in food habits, but not prized owing to poor flavor, excessively bony flesh, or inappropriate behaviour for anglers.

Fishery 1. A system to produce fish and fish-related benefits (and to minimize losses and disproducts) from the structure, dynamics and relations of the aquatic biota, the aquatic and adjacent habitat (the watersheds), and the human users of the resource.

2. The licensed premises upon which breeding, hatching, or fish-rearing facilities are situated.

Fledgling A young, fully-feathered bird in the nest.

Flood An overflow, high stream flow, or inundation that comes from a river or other body of water and causes or threatens to cause property damage.

Flood Basin The flat land area actually covered by water during the highest known floods. It is between the valley sides and the river banks and commonly contains heavy soils and limited or swampy vegetation.

Flood Frequency A relationship showing the probability that floods of a certain magnitude are equaled or exceeded in any year; or a similar relationship between flood magnitude and frequency of such magnitude being exceeded.

Flood Hydrograph A continuous record of stage and discharge, over time, produced only for the periods when the water elevation exceeds the chosen base.

Flood Peak The maximum volume of flow attained at a given point during a flood event.

Flood Plain That portion of a stream valley, adjacent to the channel, which is built of sediments during the present regimen (i.e., the state of equilibrium between a stream and its ability to transport sediments) of the stream and which is covered with water when the stream overflows its banks at flood stages. No simple, absolute flood plain commonly exists. As a consequence, flood plains are delineated in terms of some specified flood size (e.g., the 50-year flood plain – the area that would be flooded by the largest stream flow that will, on the average, occur once within a 50-year period). The largest, absolute flood plain that is ever likely to occur is sometimes referred to as the flood basin. The minimum area included is that subject to a one- percent (100-year recurrence) or greater chance of flooding in any given year.

Flood Stage The water surface elevation above some chosen reference level at which overflow of the natural banks of a stream begins to cause damage in the area.

Flood, Annual The highest peak discharge for a stream in a water year.

Flora The total plant life of an area at a time. "Vegetation" is more limited, usually meaning the large vascular plants.

Flow Currency In systems ecology, the particular form of matter or energy that flows between ecosystem components. Common flow currencies are carbon, nitrogen, phosphorus, water, heat, and, in economic systems, dollars.

Flow Diagram A technical, working diagram to show the structure of a system in terms of the flow of matter or energy, information, state variables, controls, etc. These diagrams may be used as information organizers and as bases for simulation models. In a multiple flow diagram, the simultaneous, often interdependent, flows of a number of different materials are shown.

Flow Duration A computed relationship that shows the percentage of time that specified daily discharges were equaled or exceeded in a stated number of complete years.

Flow Duration Curve (Duration Curve) A graphical representation of the number of times given stream flow volumes are equaled or exceeded during a certain period of record.

Fluctuation Irregular variation in characteristics of a population; variations in number of animals over time with irregular amplitude and/or period.

Flushout An accumulation of water that is removed suddenly from cities or surface-mined lands by heavy precipitation. Such water may contain contaminants in sufficient concentration to result in pollution of the receiving stream.

Fluvial Living in, or pertaining to, or produced by streams and flowing waters.

Flyway Routes regularly used by several species of migrating birds (4 North-South routes in the US).

Fog A cloud on the ground.

Foilage Growth of leaves on a tree or other plants.

Folkway A mode of thinking, feeling or acting common to a people or to a social group-especially a social habit that has not been rationalized or given ethical force.

Food Chain A series of spatially associated species, having feeding relations, each of which lives (at least in part) as a predator, parasite, or absorber of the next lower down in the series. "Food chains" are not isolated sequences but are interconnected. Such a pattern is called the food web.

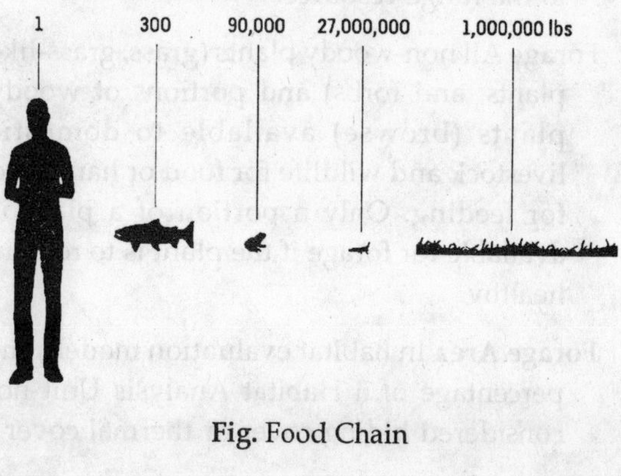

Fig. Food Chain

Food Web Feeding relationships between all the species occurring in a community, showing where each species obtains its food and what other organisms derive their food from it.

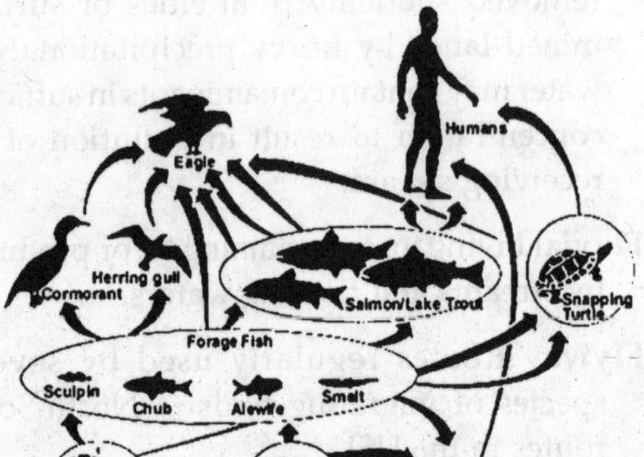

Fig. Food Web

Forage Acre An acre with a theoretical 100 percent cover of forage. A concept used to estimate grazing capacity.

Forage Acre Requirement The amount of forage, expressed in forage acres or fractions thereof, needed to sustain a mature grazing animal for a specific period without injury to the range resource.

Forage All non-woody plants (grass, grass-like plants, and forbs) and portions of woody plants (browse) available to domestic livestock and wildlife for food or harvested for feeding. Only a portion of a plant is available for forage if the plant is to remain healthy.

Forage Area In habitat evaluation models, the percentage of a Habitat Analysis Unit not considered hiding cover or thermal cover.

Forage Production (Forage Yield) The weight of forage that is produced within a designated period of time on a given area. The weight may be expressed as green, air dry, or oven dry. The term may also be modified as to time of production, e.g., annual.

Forage, Green Range forage consisting of all available and palatable live plant material and dead tips of growing grasses.

Forage, Unavailable That which livestock cannot graze (for any reason).

Forb Any herbaceous plant other than the graminoids (those in the Gramineae (true grasses), Cyperaceae (sedges) and Juncaceae (rushes) families), i.e., any non-grass-like plant having little or no woody material within it.

Forecast A prediction of future conditions and occurrences based on the perceived functioning of a system. A "forecast" differs from a projection, which is also a prediction of future conditions and occurrences, but a projection is based on an extrapolation (usually graphical) of past trends.

Foreground (Visual Distance Zone) - That part of a scene, landscape, etc., which is near the viewer (between the observer and up to 1/4 or 1/2 mile distant). The surface patterns or objects and visual elements are important in the "foreground" portions of views. In the foreground, observers are able to relate themselves to the size of individual visual elements (such as a tree) and are able to begin to sense their scale relationship to the landscape. Aerial perspective is absent or insignificant in the "foreground", and the intensity and lightness or darkness of colours are seen in maximum contrast. Wind motion in trees or on grass can be seen in this close-in area but seldom beyond.

Forest 1. A community dominated by dense, extensive tree cover.

2. An area managed for producing timber and other products, or maintained under woody vegetation for such indirect benefits as protection of watersheds or recreation.

3. An area within an administrative boundary of a forest agency, whether having trees or not.

4. Land primarily devoted to trees.

Forest Fire Any fire on forest land which is not being used as a tool in forest protection or management in accordance with an authorized plan.

Forest Highway (Legal Definition) A forest road under the jurisdiction of and maintained by a public authority and open to public travel.

Forest Land Land at least 10-percent occupied by forest trees of any size (25-percent tree canopy cover) or formerly having had such tree cover and not currently developed for non-forest use. It includes land formerly having had such tree cover (e.g., recently burned or harvested) and that will be naturally or artifically reforested. Lands developed for ***non-forest use*** include area for crops, improved pasture, residential or administrative areas, improved roads of any width, adjoining road clearing, and power-line clearing of any width. Minimiium requirement is 1 acre of at least 100 feet width.

Forest Land Use Plan A plan that gathers and coordinates the direction to be followed in the overall management of a forest. Included in plans are applicable national and regional management direction. Planning area guides, and management zone descriptions (optional), forest coordinating requirements, basic assumptions, the forest situation statement, planning unit identification and priority for study determination, unit plans and maps, and other necessary graphics.

Forest Land, Capable Forest land which is capable of growing industrial crops of wood at or above a minimum biological growth determined from research. This classification includes accessible and inaccessible, stocked and non-stocked land. Typically annual growth of 20 cubic feet per acre per year or greater.

Forest Land, Capable-Deferred Capable forestland that has been legislatively designated or administratively designated not to be used awaiting further decision (e.g., inclusion in a wilderness area).

Forest Land, Capable-Reserved Capable forestland which has been legislatively withdrawn or administratively withdrawn from timber production on a permanent basis (e.g., classified as a Wilderness Area).

Forest Land, Non-capable Forest land which is not capable of growing industrial crops of wood at least at the minimum biological growth potential of 20 cubic feet per acre annually. Forestland non-capable is classed as land not suitable for timber production.

Forest Land, Suitable Land that is to be managed for timber production on a regulated basis.

Forest Land, Unsuitable Forest land that is not managed for timber production because 1. the land has been withdrawn by Congress, the Secretary, or the Chief;.

2. The land is not producing or capable of producing crops of industrial wood;.

3. Technology is not available to prevent irreversible damage to soils, productivity, or watershed conditions;.

4. There is no reasonable assurance that lands can be adequately restocked within 5 years after final harvest, based on experience;.

5. There is at present, a lack of adequate information to respond to timber management activities; or.

6. Timber management is inconsistent with or not cost efficient in meeting the management requirements and multiple-use objectives specified in forest plans.

Forest Management 1. Giving the forest the proper care so that it remains healthy and vigorous and provides the products and amenities the landowner desires.

2. Technical definition: The application of technical forestry principles and practices and business techniques (e.g., accounting, B/C analyses) to the management of a forest.

Forest Plan A long-range plan for managing designated forest lands. This plan will provide management direction for all management programs and practices, resource uses, and resource protection measures on these lands.

Forest Range or Grazable Woodland Forest land which naturally or through silvicultural actions will support, at one or more stages in secondary succession, an herbaceous or shrubby understory that provides forage suitable for grazing or browsing animals.

Forest System Roads Roads that are part of the forest development transportation system, which includes all existing roads, as well as other special and terminal facilities designated as forest development transportation facilities-.

Forest Type Groups of stands of similar character and species composition, due to given ecological factors, differentiated from other groups of stands. The type is named based on the dominant, usually commercial, tree species existing in stands.

Forestation Establishing forests naturally or artificially on areas where it is now absent or insufficient.

Forested Forage Sometimes used in habitat evaluation models to describe forage area within forest stands that are neither hiding cover nor thermal cover. Valuable forage is often found in defined cover areas.

Forested Wetland An area characterized by woody vegetation over 20 feet tall where soil is at least periodically saturated with or covered by water.

Forestry The total system of managing and using for human benefit the natural resources that occur on and in association with land with trees.

Forest-Wide Management Requirements A set of statements which defines or indicates acceptable norms, specifications, or quality that must be met when accomplishing an activity or practice under a given set of conditions on a forest.

Format Statement A set of computer language statements used to modify the standard format of output of a computer.

Fossil Hardened remains or traces of plant or animal life from a previous geological period preserved in the Earth's crust.

Fossil fuel Any hydrocarbon deposit that can be burned for heat or power, such as petroleum, coal, and natural gas.

Fragile Lands Those land or water areas containing ecosystems, possibly but not necessarily rare, that are sensitive to external stimuli which may disturb their normal functions, resulting in system change or loss.

Fragile or Historic Lands Areas where uncontrolled or incompatible development could result in irreversible damage to important historic, cultural, scientific, or aesthetic values or natural systems which are of more than local significance. These lands include shores of rivers, lakes, and streams; rare or valuable ecosystems and geological formations; significant wildlife habitats; and unique scenic or historic areas. This is one of the three types of areas of critical environmental concern. The other types are renewable resource lands and natural hazard lands.

Fragility (Environmental Fragility) A quality or state of being expressing susceptibility to ill health or harm; the relative ability of resources to tolerate sustained use without degradation of the resource base (i.e., lowering productivity or desirable resource quality). "Fragility" applies to those resource types or locations that are more or less susceptible to degradation under the presence of whatever use categories have been considered in the evaluation process. For example, areas of vegetation-covered sand dunes may be characterized as being "environmentally fragile" for off-road vehicle uses, whereas beaches or un-vegetated dunes may be said to be "environmentally durable" for the same uses.

Frame A single image or picture. A single complete vertical scan of the cathode ray tube (CRT).

Free radicals Atomic or molecular species with unpaired electrons or an otherwise open shell configuration, usually very reactive. Specific to atmospheric chemistry free radicals are: short-lived, highly reactive, intermediate species produced by dissociation of the source molecules by solar ultraviolet radiation or by reactions with other stratospheric constituents. Free radicals are the key to intermediate species in many important stratospheric chain reactions in which an ozone molecule is destroyed and the radical is regenerated. See ozone.

Free Range Range open to grazing regardless of ownership and without payment of fees; not to be confused with open range.

Frequency (F) Number of cycles and parts of cycles completed per second. F=1/T, where T is the length of one cycle in seconds.

Frequency division multiplexing The combining of a number of signals to share a medium by dividing it into different frequency bands for each signal.

Frequency Index Any measure of the likelihood of encounter of a particular species during a specified sampling procedure.

Frequency modulation (FM) The instantaneous variation of the frequency of a carrier wave in response to changes in the amplitude of a modulating signal. As applied to APT, the radio signal from the satellite is broadcast on an FM transmitter and received on the ground on an FM radio receiver.

Frequency, Species The proportion or probability of samples in which the species occurs.

Frilling Killing trees by making a series of cuts around the bole and applying an herbicide to the wound. Frilling or girdling may be used to reduce the density of a stand, kill individual undesireable trees, or create snags or dead and down logs for wildlife.

Front A boundary between two different air masses. The difference between two air masses sometimes is unnoticeable. But when the colliding air masses have very different temperatures and amounts of water in them, turbulent weather can erupt.

A cold front occurs when a cold air mass moves into an area occupied by a warmer air mass. Moving at an average speed of about 20 mph, the heavier cold air moves in a wedge shape along the ground. Cold fronts bring lower temperatures and can create narrow bands of violent thunderstorms. In North America, cold fronts form on the eastern edges of high pressure systems. A warm front occurs when a warm air mass moves into an area occupied by a colder air mass. The warm air is lighter; 50 it flows up the slope of the cold air below it. Warm fronts usually form on the eastern sides of low pressure systems, create wide areas of clouds and rain, and move at an average speed of 15 mph.

When a cold front follows and then overtakes a warm front (warm fronts move more slowly than cold fronts) lifting the warm air off the ground, an occluded front forms.

Frost Water condensation occurring on surfaces below freezing. Condensing water turns to ice.

Frost Heave Pushing up of a soil surface by the accumulation of ice in the soil.

Fuel Any material that will sustain a wildland fire.

Fuel Break A strategically located strip of land, normally 100 to 400 feet wide, where fuel has been reduced or modified; used as a safe location from which fire fighters can attack and control a fire.

Fuel Management The practice of planning and executing treatment or control of any vegetative material which adversely affects meeting fire management direction based upon resource management goals and objectives.

Fuel Treatment A rearrangement or disposal of natural or activity fuels to reduce the fire hazard. Natural fuels are living and dead vegetation materials consumable by fire.

Fuels Wildland vegetation materials that can burn. While usually referring to above ground living and dead wildland surface vegetation, roots and organic soils such as peat are often included. In rural settings, anything that can burn.

Full Service Level A level of recreation operations, administration, and maintenance under which resources and facilities are fully protected and the recreation experience of visitors is enhanced.

Full Use The maximum use that can be made of a range during a grazing season without inducing a downward trend in range condition.

Functional Cover Hiding cover allows animal to use areas for bedding, foraging, thermal relief, wallowing, and other functions year-round. Hiding cover may contribute to security at any time, but it does not necessarily provide security during the hunting season. Hiding cover is a requisite of animal habitat and a component of security. Hiding cover alone does not provide security during the hunting season.

Functional Planning (Activity Planning, System Planning) Establishing the specific management actions to be taken in managing a single or separate resource or activity. This is undertaken only to provide details for implementing a general plan involving multiple resources and activities, e.g., it may include production scheduling or transportation plans, any one element of a comprehensive plan. Plans that do not consider all elements of a comprehensive plan may be considered as functional plans. Functional plans are sometimes referred to as comprehensive for the particular element involved, such as a "comprehensive open space plan.".

Fungi Plants that lack chlorophyll and derive their nourishment directly from other organisms (parasitic fungi) or from dead organic matter (saprophytic fungi). They include molds, yeasts, mildews, rusts, and mushrooms.

Furbearer Any mammal species or individual sought for its pelt or fur.

Fusain Fossil charcoal suggesting past presence of grass or forest fires.

Futures Unlike a scenario which is constructed for examining specific causal relationships or to aid in making a specific decision, in planning, a "future" is an all-inclusive description of what it is believed the planning region will be like if a possible plan or course of action is followed. Their use is to try to understand where existing trends will lead if they continue. Once the trends are understood, it is possible to use them to move in a number of desired directions, as in the analoguey to a sailboat and wind. In order to create a vision of the future of a region it is necessary to understand the forces that are working on that present system. To fail to understand these trends is to fail to understand the options that are open to a region.

G

Gabion A wire mesh basket filled with rocks that is used to protect erodible stream-banks or to create dams, deflectors, or other in-stream structures.

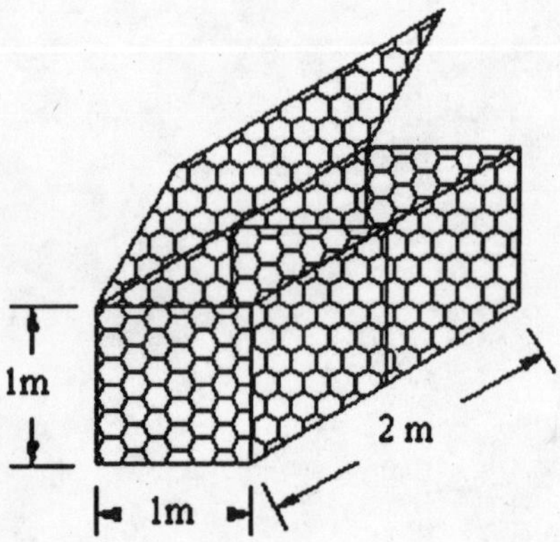

Fig. Gabion

Gain The increase in signal power produced by an amplifier; usually expressed in decibels as the ratio of the output to the input. A measure of the effectiveness of a directional antenna as compared to a non-directional antenna.

Gala hypothesis The hypothesis that the Earth's atmosphere, biosphere, and its living organisms behave as a single system striving to maintain a stability that is conducive to the existence of life.

Gall Pronounced swelling or outgrowth on a plant, usually insect or disease caused.

Game Management Unit An administrative unit for land, typically with an emphasis on management and hunting of wild game.

Game Species Any species of wildlife or fish for which hunting seasons, bag limits, and other laws and regulations have been prescribed, and which are normally harvested by hunters, trappers, poachers, and anglers under state and central laws, codes, and regulation.

Game Theory The study of decision problems in competitive situations. Game theory is the procedure for analyzing and deriving rules for making decisions when two or more people or organizations are competing for some objective.

Gamma Diversity Any of a variety of definitions of diversity relating to differences among the communities present or species present in a specific large area (e.g., a watershed).

Gap Analysis A procedure for comparing the actual distribution of faunal and floral species and vegetation classes to areas preserved or managed for their protection. Species and classes that are poorly represented in protected areas suggest gaps in the protection "for biodiversity." Areas inhabited by such species and classes thus become candidates for acquisition or designated protection.

General Administration overhead costs of agency operation and forestry (etc.) sales within an agency including salaries of rangers, supervisors, and other officials.

General Systems Theory A set of logical deductions about the behaviour of a class of objects called systems. General systems theory has been developed in order to explain and explore the general behaviour that is common to all or most apparently different empirical systems found within different disciplines. It is interdisciplinary in nature and finds a place between the level of complete generalization of mathematics and the level of specific relations within each academic field.

Geodesy A branch of applied mathematics concerned with measuring the shape of the Earth and describing variations in the Earth's gravity field.

Geographic Information System (GIS) A system for archiving, retrieving, and manipulating data that has been stored and indexed according to the geographic coordinates of its elements. The system generally can utilize a variety of data types, such as imagery maps, tables, etc.

Geold A surface of constant gravitational potential around the Earth-an averaged surface per pendicular to the force of gravity.

Geologic Materials Any of the naturally occurring mineral or rock substances. Especially used with reference to classes of such substances the properties of which are so significantly different that knowledge of their distribution is important in land use planning decisions.

Geological Area A unit of land with outstanding formations or unique geologic features of the Earth's development, including caves and fossils, subsidence areas, and earthquake-prone areas.

Geomorphic Processes The physical and chemical means which cause a modification of the Earth's surficial form, e.g., volcanic activity, weathering, mass-wasting, erosion, sedimentation, glacial action, running water, etc.

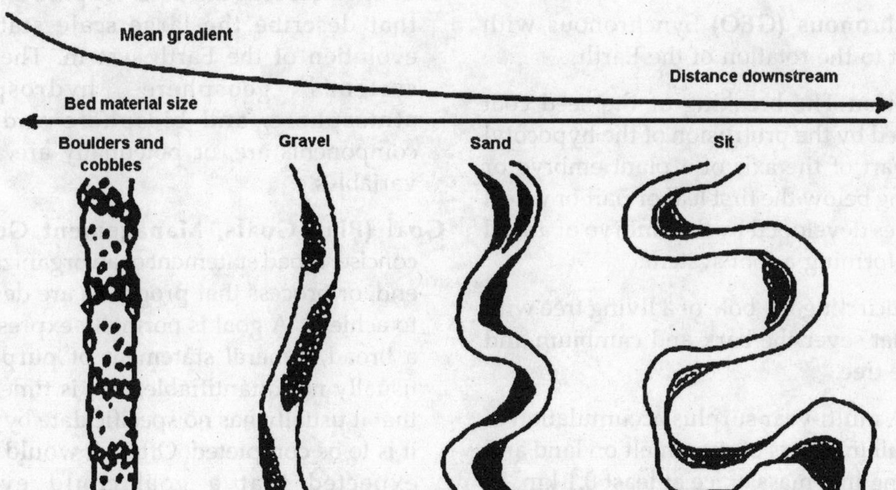

Fig. Geomorphic Processes

Geomorphology The science that deals with land and submarine relief features of the Earth's surface and seeks a genetic interpretation of them using the principles of physiography in its descriptive aspects and of dynamic and structural geology in its explanatory phases.

Geosphere The physical elements of the Earth's surface, crust, and interior.

Geostationary Describes an orbit in which a satellite is always in the same position (appears stationary) with respect to the rotating Earth. The satellite travels around the Earth in the same direction, at an altitude of approximately 35,790km (22,240 statute miles) because that produces an orbital period equal to the period of rotation of the Earth (actually 23 hours, 56 minutes, 04.09 seconds). A worldwide network of operational geostationary meteorological satellites provides visible and infrared images of Earths' surface and atmosphere. The satellite systems include the U.S. GOES, METEOSAT (launched by the European Space Agency and operated by the European Weather Satellite Organization-EUMETSAT), the Japanese GMS, and most commercial, telecommunications satellites.

Geosynchronous (GEO) Synchronous with respect to the rotation of the Earth.

Germination The breaking of the seed coat followed by the protrusion of the hypocotyl (that part of the axis of a plant embryo or seedling below the first leaf or pair or whorl of leaves developed by the embryo of a seed plant) forming a root system.

Girdle Encircling the bole of a living tree with cuts that sever the bark and cambium and kill the tree.

Glacier A multi-year surplus accumulation of snowfall in excess of snowmelt on land and resulting in a mass of ice at least 0.1 km^2 in area that shows some evidence of movement in response to gravity. A glacier may terminate on land or in water. Glacier ice is the largest reservoir of fresh water on Earth, and second only to the oceans as the largest reservoir of total water Glaciers are found on every continent except Australia.

The Global Change Research Program coordinates and guides the efforts of central agencies. The program examines such questions as, is the Earth experiencing global warming? Is the depletion of the ozone layer expanding? How do we determine and understand the causes of global climate changes? Are they reversible? What are the implications for human needs and activities?.

Global measurement All of the activities required to specify a global variable, such as ozone. These activities range from data acquisition to the generation of a data-analysis product, and include estimates of the uncertainties in that product. A global measurement often will consist of a combination of observations from a spacecraft instrument (required for global coverage) and measurements in situ (needed to provide reference points for long-term accuracy).

Global variables Functions of space and time that describe the large scale state and evolution of the Earth system. The Earth system's geosphere, hydrosphere, atmosphere, and biosphere and their components are, or potentially are, global variables.

Goal (Plan Goals, Management Goal) A concise, broad statement of an organization's end, or process that programs are designed to achieve. A goal is normally expressed as a broad, general statement of purpose, is usually not quantifiable, and is timeless in that it usually has no specific date by which it is to be completed. Often, it would not be expected that a goal could ever be completely achieved. The goal is the principal statement from which objectives must be developed. Type I-General, in Giles' 7 types of objectives.) For example, Department of Agriculture goals relating to problems associated with agricultural production are 1. To work for disease and pest control and.

2. To conduct research on production. Goals are normally stated in terms of the fulfilment of broad public needs, the preservation of fundamental constitutional principles, the achievement of targeted levels of excellence, the alleviation of major problems, or other purposes to be served by government effort. A goal usually has a societal consensus.

Goal Programming A mathematical approach to optimization. A series of goals or objectives are established which can be expressed as mathematical equations. However, due to constraints on the original problem, not all the goals can be met simultaneously. It is necessary to allow positive and negative slack within these goals equations. This slack is allocated to "slack variables," which are combined with the goals to form 1. A linear equation to be minimized or maximized and.

2. A series of accompanying constraint equations. This system of equations is then solved by conventional linear programming techniques.

Goods and Services The various outputs, including on-site uses, produced from wildland resources.

Gouging Gouging is an act to produce a land surface configuration using mechanical reclamation equipment. It is intended to trap precipitation, increase infiltration, and reduce erosion.

Grade 1. The inclination or slope of a stream channel or ground surface, usually expressed in terms of the ratio or percentage of number of units of vertical rise or fall per unit of horizontal distance.

2. The finished surface of a roadbed, top of an embankment or bottom of an excavation.

3. To establish a new land surface profile.

Graphic Scale A bar, or other such graphic device, on a scale drawing, divided into properly scaled-down measurement units. In subsequent photographic reduction or enlargement of a drawing, the "graphic scale" becomes a device from which distances can be readily and accurately measured.

Grayscale Environmental satellite scanners, rather than photographing a scene, scan a scene line-byline measuring light or heat levels and transmitting this information as a video image via an amplitude modulated (AM) subcarrier contained in the satellite's FM signal. The video image-a 2400 Hz tone-is amplitude modulated to correspond to the light and dark areas sensed, with the louder portion of the tone representing the lighter areas of the.

Grazing Consumption of range or artificial pasture forage by animals.

Grazing Capacity, Maximum The maximum number of animals that a given range area is capable of supporting without inducing injury to soil, vegetation, or a watershed.

Grazing Land, Native Grazing land used primarily for producing of native forage plants maintained or manipulated primarily through grazing management. Native grazing lands include rangeland, woodland that may be properly grazed, and native pasture, individually or collectively.

Grazing Management Plan - A program of action designed to secure the best practicable use of range. (Objectives are critical for determining "best" and "practicable.").

Grazing Permit (Grazing License) Official, written permission to graze a specified number, kind, and class of livestock for a specific period on a defined range allotment.

Grazing Preference Probable selection of certain plants over others by grazing animals or in administrating public lands, the basis upon which permits and licenses are issued for grazing use.

Grazing Pressure The actual animal-to-forage ratio at a specific time, e.g., three animal units per ton of standing forage.

Grazing Privilege Permissive use of lands for grazing by livestock.

Grazing Right A right to graze land, permanently vested in the beneficiary as specified by the terms of the law, land title, or contract.

Grazing Season A period of grazing within which optimum use of the forage resource may be obtained.

Grazing System The manipulation of livestock grazing to accomplish a desired result. A planned program of livestock management to accomplish pre-determined objectives. A system selected for an area that can be integrated with the stockperson's operation and provides for coordinating range use with other resource uses.

Grazing Trespass The grazing of livestock on a range area without proper authority.

Grazing, Continuous Allowing domestic livestock to graze a specific area throughout the grazing season; not necessarily synonymous with grazing for a 12-month period or a calendar year.

Grazing, Deferred Discontinuance of grazing by livestock on an area for a specified period of time during the growing season to promote plant reproduction, establish new plants, or restore plant vigour.

Grazing, Deferred Rotation Discontinuance of grazing on various parts of a range in succeeding years, allowing each part to rest successively during the growing season to permit seed production, establish seedlings, or restore plant vigour. Two, but usually three or more, separate grazing units are required. Control is usually insured by unit fencing, but may be obtained by herding on sheep ranges.

Grazing, Heavy A comparative term (often used in research but may apply to practical management as well) which indicates that the stocking rate of a pasture is relatively greater than other pastures. Often erroneously used to mean over use.

Grazing, Intermittent Grazing range or artificial pasture for indefinite periods at indefinite intervals.

Grazing, Light A comparative term (often used in research but may apply to practical management as well) which indicates that the stocking rate of one pasture is relatively less than that of other pastures. Often erroneously used to mean "under-use" of grazing capacity.

Grazing, Moderate A comparative term (often used in research but may apply to practical management as well) which indicates that the stocking rate of a pasture is in between the rates of that on other pastures.

Grazing, Premature Grazing before range plants have reached sufficient seasonal development to avoid over-use or before the soil has become firm enough to withstand the presence of grazing animals. Range readiness is reached when these conditions are met and range utilization may properly begin.

Grazing, Rest Rotation An intensive system of management whereby grazing is deferred on various parts of the range during succeeding years, allowing the deferred part complete rest for one year. Two or more units are required. Control by fencing is usually necessary on cattle range, but may be obtained by herding on sheep ranges.

Grazing, Rotation Grazing two or more pastures or parts of a range in regular order, with definite recovery periods between grazing periods. Where only two fields are involved, sometimes called alternate grazing.

Grazing, Selective The grazing of certain plant species on the range to the exclusion of others.

Grazing, Spot Repeated grazing of small areas while adjacent areas are lightly grazed or unused.

Green Belt Open space areas around villages, towns or cities in which development is strictly controlled and (usually not permitted). A useful device for preventing the sprawl of towns or the union of one town with another, providing townspeople with a welcome release from the strain of city life, and protecting the special character of some towns. The main deficiency of green belt preservation is that it seems to encourage development leap-frogging over the green belt to the land beyond it.

Greenhouse effect Process by which significant changes in the chemistry of Earth's atmosphere may enhance the natural process that warms our planet and elevates temperatures. If the effect is intensified and Earth's average temperatures change, a number of plant and animal species could be threatened with extinction.

Certain gaseous components of the atmosphere, called greenhouse gases, transmit the visible portion of solar radiation but absorb specific spectral bands of thermal radiation emitted by the Earth. The theory is that terrain absorbs radiation, heats up, and emits longer wavelength thermal radiation that is prevented from escaping into space by the blanket of carbon dioxide and other greenhouse gases in the atmosphere. As a result, the climate warms. Because atmospheric and oceanic circulations play a central role in the climate of the Earth, improving our knowledge about their interaction becomes essential.

Greenhouse gas A gaseous component of the atmosphere contributing to the greenhouse effect. Greenhouse gases are transparent to certain wavlengths of the suns' radiant energy allowing them to penetrate deep into the atmosphere or all the way into the Earths' surface. Greenhouse gases and clouds prevent some of infrared radiation from escaping, trapping the heat near the Earths' surface where it warms the lower atmosphere. Alteration of this natural barrier of atmospheric gases can raise or lower the mean global temperature of the Earth.

Greenhouse gases include carbon dioxide. methane, nitrous oxide, chloro fluorocarbons, and water vapour Carbon dioxide, methane, and nitrous oxide have significant natural and human sources while only industries produce chloro-fluorocarbons. Water vapour has the largest greenhouse effect, but its concentration in the troposphere is determined within the climate system. Water vapour will increase in response to global warming, which in turn may further enhance global warming.

Greentree Reservoir A forested wetland inundated during the dormant period of tree growth to provide temporarily aquatic habitat, usually for waterfowl, without damaging tree survival.

Gross feature map Map that displays geographic characteristics rather than political boundaries.

Ground control (points) Identifiable points on the ground whose locations on the surface of the Earth are accurately known for use as geodetic references in mapping, charting, and other related mensuration applications.

Ground Water Subsurface water occupying the saturation zone (where all openings in soils and rocks are filled), from which wells and springs are fed. The upper surface of this zone of saturation forms the "water table.".

Group Cutting (Group Selection Cuts) A silvicultural system of uneven-age management in which the logging operation removes all merchantable timber from an area smaller than that normally recognized in timber type and condition class mapping. It removes small groups of trees to meet a predetermined goal of size, distribution, and species in their remaining stands. Trees are removed in small groups at a time to improve the stand and encourage regeneration in an uneven-aged forest. Group cuts end up being one-fifth (0.2) to 1.0 acre in area, but area is not a criterion for selecting trees for removal. Group, patch, strip, and stand cutting, however, have been used as size designations of the same logging approach, commonly known as clear-cutting. All require the removal of all merchantable trees except a few which may sometimes be carried over where needed for other forest values such as nesting sites, food caches, etc. Group cuts are made on areas that are smaller than the size normally recognized in type and condition class mapping. Patch cuts are made on areas that are of a size generally mapped for type and condition and for control, but which do not include the entire stand of which they are a part. Strip cuts are on areas running through a stand and usually of a width equal to one to two times the general stand height. Stand cuts are on areas large enough to be practical for management.

Growing Stock Level Timber growing on any area expressed in either stems per acre or square feet of basal area.

Growth Rate With reference to wood, the rate at which the wood substance has been added to the tree at any particular point; usually expressed in terms of number of rings per inch. Growth rate bears an inverse relationship to number of rings per inch. Also applies to volume, value, or other types of increase in trees or stands.

Growth Rings The layers of wood laid down each growing season, also called annual rings. These rings frequently are visible when a tree is cut and may be used to estimate the age of the tree, as well as to determine the rate of its growth.

Guided wave Electromagnetic or acoustic wave that is constrained within certain boundaries, as in a wave guide (transmission line).

Guideline A steering or usual course of action, but neither mandatory nor requiring a superior's approval for deviation from it. A statement, information and guidance, which assists in decision making and determining the course of direction to be taken in any planned project and action to accomplish or achieve specific desired conditions and objectives.

Gulf A large arm of an ocean or sea extending into a land mass.

Gulf stream A warm, swift ocean current that flows along the coast of the Eastern United States and makes Ireland, Great Britain, and the Scandinavian countries warmer than they would be otherwise.

Gully A channel or miniature valley cut by concentrated runoff but through which water commonly flows only during and immediately after heavy rains or during the melting of snow. The distinction between "gully" and rill is one of depth. A gully is sufficiently deep that it would not be oblitreated by normal tillage operations, whereas a rill would be smoothed by ordinary farm tillage.

H

Habitat The area or region where a particular type of plant or animal lives and grows. 1. The total elements in an organism's physical surroundings which have a direct bearing on that organism's function and survival.

2. The total of environmental conditions of a specific place that is occupied by an organism, by a population, or a community of interest.

3. The "home" or place where animals live, reproduce, and die.

4. A geographical niche.

5. The natural or native environment of a plant or animal.

Habitat Analysis Unit An area of land defined by geographic or administrative means and selected as the unit for evaluating the quality of habitat of a species, guild, or other taxon.

Habitat Capability Model A model which depicts the relationship of a species to a variety of habitat factors which provide for quantitative predictions of a species response (animal abundance numbers) to habitat change.

Habitat Capability The capacity of a given area, nearly equivalent to carrying capacity. The estimated ability of an area, given existing or predicted habitat conditions, to support a wildlife, fish, or plant population.

Habitat Diversity The mix of habitats; or size variety; or numbers of layers present; or number of types; or evenness of stand sizes and shapes by type. Also relative rarity. Not always lack of similarity. Occasionally differences in an area observed over time (instability).

Habitat Effectiveness Percentage of available habitat that is usable by wildlife outside the hunting season. Originating in the road density models as a means of expressing habitat loss associated with open forest roads, it has since been used to express habitat quality, hunting season security, habitat capability, carrying capacity, and several other conditions.

Habitat Management Unit (HMU) A unit of land approximately 4,000 acres in size, the boundaries of which fall on compartment boundaries, and which contains a mix of habitat types. At least one of these types must be a pond or stream with wetland potential.

Habitat Type A small unit of land from a few to over 100 acres lying within a given climatic mineralogical zone and supporting a probable sequence of vegetation growing on a unique type of soil material.

Hail Precipitation composed of balls or irregular lumps of ice. Hail is produced when large frozen raindrops, or almost any particles, in cumulonimbus clouds act as embryos that grow by accumulating supercooled liquid droplets. Violent updrafts in the cloud carry the particles in freezing air; allowing the frozen core to accumulate more ice. When the piece of hail becomes too heavy to be carried by upsurging air currents it falls to the ground.

Hard Mast Fruits of oaks, hickories, pines, and beech trees that are important foods of many species of wildlife in the fall and winter.

Hardness A property of water that may cause formation of an insoluble residue when used with soap and causing formation of a scale in vessels in which water has been allowed to evaporate. It is primarily due to the presence of calcium and magnesium ions. It is generally expressed in milligrams per litre.

Hardpan A hardened soil layer in the lower A or in the B horizon caused by cementation of soil particles with organic matter or with materials such as silica or calcium carbonate. The hardness does not change appreciably with changes in moisture content, and pieces of the hard layer do not disintegrate in water.

Hardware The electrical and mechanical components of a system, as opposed to software.

Hardwood A broad-leaved flowering tree, usually deciduous, as distinguished from a conifer. Trees belonging to the botanical group of angiosperms; trees such as oaks, maples, ashes, elms. It does not necessarily refer to the hardness of the wood.

Harvest Cutting Method A means of removing all or part of a stand of timber, either by a single cutting or a series of regeneration cuttings.

Harvesting A loose term for removing produce (i.e., a merchantable material) from the forest for utilization; cutting, sometimes further initial processing (e.g., with trees, topping and trimming) and extraction (removal from the forest).

Haul Road Road from pit to loading dock, tipple, ramp, or preparation plant used for transporting mined or other material by truck.

Haze Fine dry or wet particles of dust, salt, or other impurities that can concentrate in a layer next to the Earth when air is stable.

Head of the Hollow (Valley-Fill Method) A method for surface-mined land in which overburden material from adjacent contour or mountain top mines is placed in compacted layers in narrow, steep-sided hollows so that surface drainage is possible.

Headwater The small rivulets that are the source of a stream or river. Mapped as starting at the last crenulation in a contour line below a ridge or crest on a topographic map.

Heartwood The inner core of a woody stem, wholly composed of nonliving cells and usually differentiated from the outer enveloping layer (sapwood) by its darker colour.

Heat Capacity Mapping Mission (HCMM) A two-channel radiometer launched by NASA to measure the thermal properties of the terrestrial surface. It had an application to identify and locate rocks and minerals. One radiometer channel was in the visible to near infrared (0.5-1.1 micrometers), and the other in the thermal infrared (10.5-12.5 micrometers). The instantaneous field of view (IEOV) was about 600 meters.

Heat Island A small area in a landscape with air temperatures measurably greater than that of its surroundings. Urban areas form heat islands. The "heat island effect" is an air circulation phenomenon generally found in and around cities. Warm air tends to concentrate in the city's center, probably because of the mass of tall buildings and concentration of paved surface. This warm air rises, carrying with it a burden of pollution. As the polluted air moves up over the city, spreads out, cools, and sinks at the city's periphery, it forms a distinctive ceiling (composed of suspended pollutants) known as a haze hood or dome. Cooler air from the edge of the city flows into the center to replace the rising air and is followed again by cool air from the city's edge. Thus, a self-contained circulatory system is set up that is altered only by a strong wind capable of breaking up the pollution-formed ceiling. Such a heat island, if it becomes stabilized for a sufficient length of time, can create hazardous concentrations of pollutants.

Heat_balance The equilibrium existing between the radiation received and emitted by a planetary system.

Heavy Metals Metals having a specific gravity (i.e., weight in comparison to weight of an equal volume of water) of 5.0 or over, and generally toxic in relatively low concentrations to plant and animal life. Such metals can persist in animal tissue and are capable of increasing in concentration as they pass upward through the food chain. Major sources of heavy metal contamination are pesticides, limestone and phosphate fertilizers, manure and sewage sludge, and mine wastes. Examples include lead, mercury, cadmium, and arsenic.

Hedging The appearance of browse plants that have been browsed so as to appear artificially clipped. The consistent browsing of terminal buds of browse species causes excessive lateral branching and a reduction in upward growth.

Heel-in To store young trees prior to plannng by placing them in a trench and covering the roots or rooting portions with soil.

Height, Merchantable The height of a tree (or length of its trunk) up to which a particular product may be obtained. For example, if the minimum usable diameter of pulpwood sticks is 4 inches, the merchantable height of a straight pine tree would be its height up to a trunk diameter of 4 inches. Note, one must know the product being cut to estimate merchantable height.

Height, Total The height of a tree from the ground level to the top of its crown.

Hemisphere Half of the Earth, usually conceived as resulting from the division of the globe into two equal parts, north and south or east and west.

Herb Any flowering plant except those developing persistent woody stems above ground.

Herbaceous Vegetation The low-growing, non-woody plants in a forest understory, including grasses (graminoids),wildflowers, and ferns.

Herbage Herbs taken collectively, usually used in the same sense as forage, except that it may include material not acceptable to grazing or browsing animals.

Herbicide (Weedicide) A chemical substance used to prevent, inhibit, or destroy non woody plant growth. If its effectiveness is restricted to a specific plant or type of plant, it is known as a selective herbicide. If its effectiveness covers a broad range of plants, it is considered to be non-selective or broad spectrum. A herbicide is one type of pesticide. In common usage, however, often used interchangeably with the words phytocide (plant killer) and silvicide (tree killer).

Herbivore An animal that feeds primarily on living plant substances.

Herd Home Range The area that a social group of ungulates traverses during normal activities. It usually includes the total range for a year.

Hertz The international unit of frequency equal to one cycle per second. Radio frequencies are usually expressed in kilohertz/kHz (1,000 cycles per second) or megahertz/MHz (1,000,000 cycles per second).

Hertzian waves Radio waves or other electromagnetic radiation resulting from the oscillations of electricity in a conductor.

Heterogeneous Differing in kind; not uniform; having qualities which are significantly different throughout; possessed of different characteristics.

Heterotrophic Organism An organism that is dependent on organic matter for food. In contrast, an autotrophic organism is capable of using inorganic substances as its source of food.

Hibernation A condition in which an animal's metabolism is purposely slowed to endure prolonged periods of adverse environmental conditions. normally several months at a time.

Hiding Cover, Functional Hiding cover allows animal to use areas for bedding, foraging, thermal relief, wallowing, and other functions year-round. Hiding cover may contribute to security at any time, but it does not necessarily provide security during the hunting season. Hiding cover is a requisite of animal habitat and a component of security. Hiding cover alone does not provide security during the hunting season.

Hiding Cover, Structural Vegetation capable of hiding 90 percent of a standing adult animal from the view of a human at a distance equal to or less than 200 feet. As a site-specific vegetative component of security, the quality of hiding cover varies inversely with sight distance.

Hierarchical Classification in which successively lower (i.e., more specific) level units must fit entirely within the separate units delineated by the next higher (i.e., less specific) level in that system. The boundaries of the more specific level units must occur within or coincide with (and not cross) those boundaries delineated for the units of the next less specific level in the classification system.

Hierarchical Structure Group(s) of things (or persons) arranged in order of rank, grade, and class.

High A digital logic state corresponding to a binary "1.".

High Grading Removing the mature, high quality trees from a stand and leaving inferior species and defective trees. "Take the best and leave the rest." Generally regarded as a poor forestry practice.

High Quality Hardwoods Hardwood trees or stands that will yield high value timber products such as face veneer, knot free lumber, furniture or specialty product stock, or flooring.

High Resolution Doppler Imager (HRDI) Carried on UARS, it measures stratospheric winds.

High-Resolution Infrared Radiation Sounder (HIRS) Instrument carried by NOAA polar-orbiting satellites that detects and measures energy emitted by the atmosphere to construct a vertical temperature profile from the Earth's surface to an altitude of about 40 km. Measurements are made in 20 spectral regions in the infrared band.

Highwall The unexcavated face of exposed overburden and coal or ore in an open-pit mine or the face or bank on the uphill side of a contour strip mine excavation.

Histopath Analysis Determines changes in an organism's tissue structure as a result of some physical (parasitism) or chemical (toxic substance) activity.

Historic and Cultural Sites Sites associated with the history, tradition or cultural heritage of national, state or local interest and of enough significance to merit preservation or restoration. The location of the feature establishes the site. Recreational activities on these sites center on sight seeing, enjoying and studying the historic or cultural feature. Management is generally limited to activities that would effect such preservation and restoration as may be necessary to protect the features from deterioration and to interpret their significance to the public. Administration is by all levels of public agencies and private landowners who identify, set aside, and manage historic and cultural areas.

Historic Preservation Protecting, rehabilitating, restoring, and reconstructing districts, sites, buildings, structures, and objects significant in history, architecture, archaeology or culture.

Historical Area Sites and areas which have been designated as containing important evidence and remains of the life and activities of early settlers and others who used or visited the area or the sites where important events took place. Examples are battlegrounds, remnants of mining camps, old cemeteries, important pioneer roads and trails, and early trading sites.

Historical Range of Variability The limits of change in composition, structure, and processes of the biological and physical components of an ecosystem resulting from natural variations in the frequency, magnitude, and patterns of natural disturbance and ecological processes characteristic of an area before European settlement. Estimates are made for a specified period of time and include the effects of pre-European settlement human activities.

Holism The doctrine that the universe-including life in all its forms and the inorganic environment – is correctly seen in terms of interacting wholes that are more than the mere sum of elementary particles.

Holistic An attempt to study and master the knowledge and control of entire systems. An emphasis on functional relations between parts and whole; a doctrine in which a whole cannot be expressed as the simple sum of its parts.

Home Range The area which an animal uses during its normal activities, not to be confused with territory.

Homeostasis 1. An important process or mechanisms of an organism (perhaps biotic community) for regulation (perhaps pre-disposed self-regulation) toward a constant or standard condition. All communities, except possibly the very simplest types, appear to contain regulatory mechanisms that enable them to adjust to the changing conditions of their physical environments.

2. The maintenance of a high degree of uniformity in functions of an organism or interactions of individuals in a population or community under changing environmental conditions. The constant function or status that results from the capabilities of organisms to make compensatory adjustments. (Negative feedback is at work.).

Homogeneous Response Unit (Response Unit) A particular land (or water) unit that is delineated on the assumption that it is sufficiently uniform within its boundaries to respond in a homogeneous manner throughout the area to any set of inputs. That is, any treatment, stimulus, or set of conditions will cause approximately the same reaction, response, or set of outputs when applied to each location within the unit.

Horizon, Soil A layer of soil approximately parallel to the land surface with more or less welldefined characteristics that have been produced through the operation of soilbuilding processes. 1. A-horizon—The upper horizon of the mineral soil, from which material has been removed by percolating waters. The horizon of eluviation. Commonly divided into a dark coloured Al horizon containing a relatively high content of organic matter, and light-coloured M horizon of maximum leaching,.

2. B-Horizon—The horizon of deposition, to which materials have been added by percolating waters, the horizon of illuviation.

3. C-Horizon—The weathered parent material.

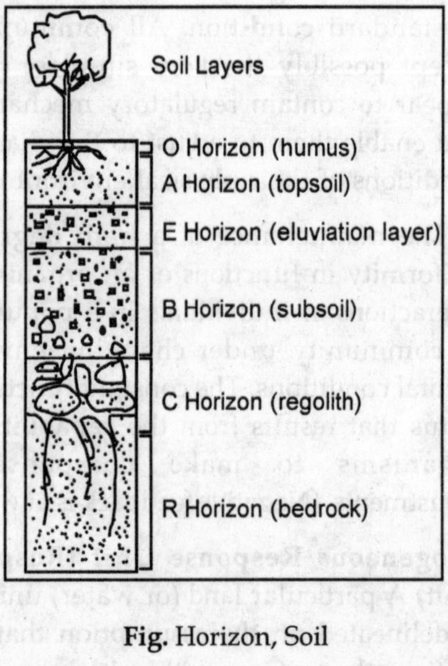

Fig. Horizon, Soil

Horizontal Diversity A comparative measure of occurrence and complexity of distribution of cover types over a wide area encompassing different cover types. Stand site, stand configuration, species composition, and a variety of age classes are some variables that make up horizontal diversity in a forested environment.

Horse latitudes The subtropical latitudes (30-35 degrees), where winds are light and weather is hot and dry. According to legend, ships traveling to the New World often stagnated in this region and had to throw dead horses overboard or eat them to survive, hence the name horse latitudes. See wind.

Human and Community Development System A system having actions that help people and communities to help themselves. The system includes activities that provide 1. Youth development through working and learning experiences,.

2. Adult employment and training opportunities through various central manpower programs,.

3. Community planning and development information and services for rural areas, and.

4. Technical forestry assistance for urban areas in the establishment, management, and protection of needed open space and the use of trees and woody shrubs. One of the six "systems" established by the Forest Service to have a systematic, orderly way to view and evaluate its many diverse but interrelated activities. It grouped its various programs into these six ".systems," each of which incorporates all the activities concerned with developing and managing a specific resource. The six "systems" are: Land and Water, Timber Resource, Outdoor Recreation and Wilderness, Rangeland Grazing, Wildlife and Fish Habitat, and "Human and Community Development".

Human Community The spatial or territorial unit in social organization in which members have a psychological cohesiveness or feeling of belonging, one not forced, and without loss of individuality.

Human Resource Unit A geographic area characterized by particular patterns of cultural lifestyle and economic conditions.

Humidity The amount of water vapour in the air. The higher the temperature, the greater the number of water molecules the air can hold. For example: at 60 oF (15°C), a cube of air one yard on each side can hold up to 4.48 ounces of water. At 104oF (40°C), the same cube of air can hold up to 17.9 ounces of water.

Relative humidity describes the amount of water in the air compared with how much the air can hold at the current temperature. Example: 50% relative humidity means the air holds half the water vapour that it is capable of holding; 100% relative humidity means the air holds all the water vapour it can. At 100% humidity no more evaporation can occur until the temperature rises, or until the water vapour leaves the air through condensation. Absolute humidity is the ratio of the mass of water vapour present in a system of moist air to the volume occupied by the mixture, that is, the density of water vapour.

Humus Layer The top portion of the soil which owes its characteristic features to its content of humus. The humus (brown or black organic molecules, gellatinous, following processing by bacteria, fungi, and soil organisms) may he incorporated or unincorporated in the mineral soil.

Hurricanes Severe tropical storms whose winds exceed 74 mph. Hurricanes originate over the tropical and subtropical North Atlantic and North Pacific oceans, where there is high humidity and light wind. These conditions prevail mostly in the summer and early tall. Since hurricanes can take days or even weeks to form, time is usually available for preventive or protective measures. From space, hurricanes look like giant pinwheels, their winds circulating around an eye that is between 5 and 25 miles in diameter. The eye remains calm with light winds and often a clear sky.

Hurricanes may move as fast as 50 mph, and can become incredibly destructive when they hit land. Although hurricanes lose power rapidly as soon as they leave the ocean, they can cause high waves and tides up to 25 feet above normal. Waves and heavy flooding cause the most deaths during a hurricane. The strongest hurricanes can cause tornadoes.

Hydraulic Conductivity The rate at which water moves through aquifer material under a unit hydraulic gradient, expressed as volume per unit time per unit cross section ($ft^3/day/ft^2$ or $M^3/day/M^2$ reduced to feet per day or meters per day).

Hydrochloroflourocarbon (HCFC) One of a class of compounds used primarily as a CFC substitute. Work on CFC alternatives began in the late 1970s after the first warnings of CFC damage to stratospheric ozone. By adding hydrogen to the chemical formulation, chemists made CFCs less stable in the lower atmosphere enabling them to break down before reaching the ozone layer. However, HCFCs do release chlorine and have contributed more to atmospheric chlorine buildup than originally predicted. Development of non-chlorine based chemical compounds as a substitute for CFCs and HCFCs continues.

Hydrograph A graph showing variation in the water depth in a stream or the volume of water flowing past a point in a stream over a period of time.

Hydrologic Cycle (Water Cycle) The pathways through which water is cycled in the terrestrial biosphere. The circuit of water movement from the atmosphere through various stages or processes on the ground (such as precipitation, interception, runoff, infiltration, percolation, storage) and then back to the atmosphere again by evaporation, and transpiration.

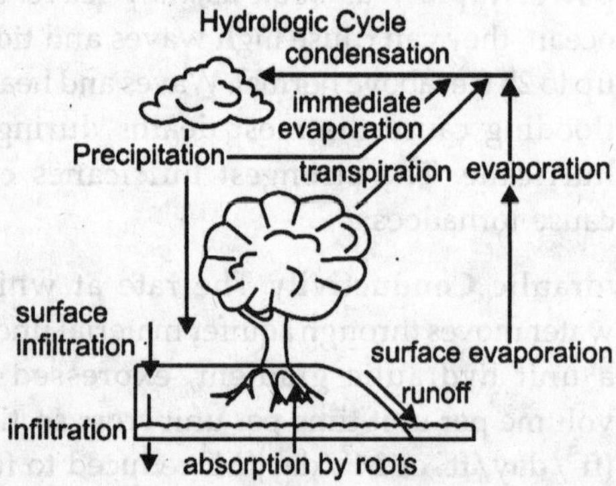

Fig. Hydrologic Cycle

Hydrology A science dealing with the properties, distribution, and circulation of water, specifically the study of water and its behaviour on the surface of land, in the soil and underlying rocks, and in the atmosphere, particularly with respect to evaporation and precipitation.

Hydrosere Stages of ecological succession in a pond environment.

Hydrosphere The totality of water encompassing the Earth, comprising all the bodies of water; ice, and water vapour in the atmosphere.

Hygric (Hydric) A habitat characterized by wet or moist conditions rather than mesic (moderate) or xeric (dry) conditions.

Hygrometer Instrument that measures water vapour content in the air and communicates changes in humidity visibly and immediately through a graph or a dial. There are three types of hygrometers:.

The hair hygrometer uses a human hair as the sensing instrument. The hair lengthens when the air is moist and contracts when the air is dry but remains unaffected by air temperature. However; the hair hygrometer cannot respond to rapid fluctuations in humidity.

An electric hygrometer uses a plate coated with carbon. Electrical resistance of the carbon coating changes as the moisture content of the air changes-changes that translate into relative humidity. This type of hygrometer is used frequently in the radiosonde.

An infrared hygrometer uses a beam of light containing two separate wave lengths to gauge atmospheric humidity. One of the wavelengths is absorbed by water vapour; the other is unaffected, providing an extremely accurate index of water vapour for paths of a few inches or thousands of feet. See psychrometer.

Hypolimnion The deep layer of a lake below the thermocline and removed from surface influences such as direct heating by sunlight and aeration by vertical circulation.

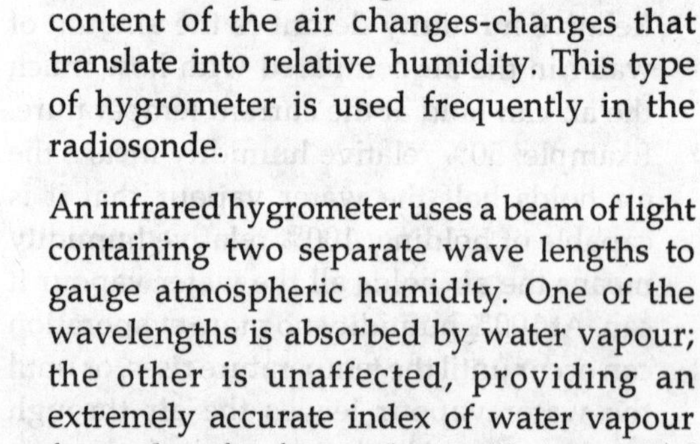

Fig. Hypolimnion

Hyporheic Zone A layer of the landscape; the saturated interstitial [volumes] beneath streams and rivers and into the banks that exchange water with the channel and surface. It contains water of both subsurface and stream channel origin.

I

I Impact, Ecological The observable or suspected effects of one or more perturbations on a factor (usually biological) of an environmental system. There are no value judgements, e.g., positive or negative impacts. An impact is simply an effect or a change of condition.

Ice shelf A thick mass of ice extending from a polar shore. The seaward edge is afloat and sometimes extends hundreds of miles into the sea.

Image Pictorial representation of data acquired by satellite systems, such as direct readout images from environmental satellites. An image is not a photograph. An image is composed of two-dimensional grids of individual picture elements (pixels). Each pixel has a numeric value that corresponds to the radiance or temperature of the specific ground area it depicts.

Impact Assessment, Environmental An evaluation and objective prediction of the environmental impacts of a proposed action using a systematic, interdisciplinary approach that integrates social and natural sciences and environmental design arts.

Impacts, Ecological, Types of 1. Accumulative - The sum of a variety of environmental changes and their total effect.

2. Acute - A person-induced ecological effect commonly resulting from a high level of disturbance but of short duration.

3. Beneficial - A person-induced ecological change which is useful to people and favourable to the system.

4. Chronic - A person-induced ecological effect commonly resulting from low levels of disturbance but often persisting over long periods of time.

5. Detrimental - A person-induced ecological effect which damages ecosystem structure or function.

6. Direct (primary impact) - Direct or primary impacts are those in which the causative agent impinges directly upon the responding ecological components, e.g., land clearing for construction causes direct loss of vegetation.

7. Indirect (secondary impact) - Indirect or secondary effects are those in which the person-caused change in the environment creates one or more intermediary effects in a chain of events leading to the impact being observed. Certain agencies (e.g., EPA) differentiate indirect and secondary impacts. For example, an indirect impact would occur when construction of a dam causes reduction

of stream flow, in turn causing elimination of riverine wetlands downstream. Secondary impacts occur when a wastewater treatment project induces urbanization that in turn has environmental impacts.

8. Long-term a. which last the lifetime of the project or past the construction phase, or for the duration of maintenance and operation of the project, e.g., visual impact due to the design and building materials of a structure. It should be noted that some long-term impacts might be reversible. The time frame of reference should be stated when "long-term" impacts are discussed.

b. "Long-term" and "short-term" are two phrases that have caused considerable confusion in the environmental assessment process since there are no definitions for time periods encompassing either short-term or long-term. Perhaps a better selection of terms would be to divide the temporal phases of a project into pre-construction and planning, construction, immediate post-construction and initial operation, operation, and post-operation. In some cases, all of the above temporal phases, even through project operation, can occur over a short time period.

9. Moderate and severe - Moderate and severe effects are gradations of adverse impacts. Moderate could be characterized as partial elimination, dislocation, impairment or alteration of biota or use of resources and facilities. Total elimination, dislocation, impairment, or alteration of biota or use of resources and facilities could characterize "Severe".

10. Short-term a. Typically construction-induced impacts which last only as long as construction, or may be characteristic non-construction activities, such as spraying, or may occur during the operation of a facility, such as brief air pollution episodes due to temperature inversions. Since no strict definition of "short-term" exists, the time frame of reference should be stated.

b. Any impact occurring over a relatively short period of time.

11. Significant - An impact resulting in measurable changes in indicator parameters, community dynamics and succession.

12. Synergistic - Impacts can have synergistic effects in that individually they may be relatively unimportant but together may form impacts which are "greater than the sum of their parts," e.g., timber harvesting practices may increase water temperatures, which make fish more susceptible to disease, while also decreasing the dissolved oxygen in the water, creating conditions of severe stress for fish.

13. Tertiary - Observed change resulting from a perturbation which directly impinges on the biota with two or more prior cause-and effect steps.

Impingement Incidents in which fish become caught and held by suction against intake screens at dams such as those utilized in power plant cooling systems.

Implementation Plan A document for giving practical effect to, or ensuring the actual fulfillment of some regulatory standard or goal by specific measures.

Impoundment A body of water formed by collecting and confining stream flow (as in a reservoir).

Imprinting A short-term rapid learning process early in life (fixing young on parents or habitat) which is genetically irreversible. More prevalent in precocial young.

Improvement Cutting Eliminating or suppressing less valuable (poor species, form, or genetics) trees to increase growth rates of more valuable trees, typically in mixed uneven-aged forest. A cutting is made in an immature stand to improve the stand composition and character.

In developing and maintaining land management plans for use on units of the

National Forest System, the Secretary shall use a systematic interdisciplinary approach to achieve integrated consideration of physical, biological, economic, and other sciences.

In situ Latin for "in original place." Refers to measurements made at the actual location of the object or material measured. Compare with remote sensing.

Incommensurable Values Resource yields for which no method yet exists for conversion to a common measurement scale or unit (such as dollars) for purposes of comparative evaluation of relative importance, e.g., comparison of the aesthetic benefits with the timber yields of land is difficult because it is often impossible to express the former objectively in dollar values or the latter in equivalent aesthetic units.

Incompatible Uses Land uses which cannot exist together by reason of either competition for limited resources or use by products which prevent the alternative use, e.g., timber harvesting and wilderness preservation are incompatible uses for one piece of land.

Increaser Plant Species (Increaser) Plant species (e.g., rangeland plants) of the original vegetation that increase in relative amount, at least for a time, under overuse.

Increment Borer An auger-like instrument wilh a hollow bit, used to extract cores from trees for determining their growth and age.

Incrementalism An alternative planning approach to comprehensive planning based on the assumption that due to practical difficulties, true comprehensive planning cannot exist. As an alternative, it is suggested that planners concentrate on a recursive process of choosing among policies based on incremental changes to existing policies without trying to set overall goals and without basing the policies on theoretical predictions of the consequences of the policies or on an understanding of the system being regulated. Instead the planners would rely heavily on the record of past experience with small policy steps to predict the consequences of similar steps extended into the future. Social scientists, politicians, or public administrators do not yet know enough about the social world to avoid repeated error in predicting the consequences of policy moves. Practicing incrementalism, a policy-maker consequently expects that his or her policies will achieve only part of what is hoped for. The policies, the same time, will produce unanticipated and undesirable consequences. If the policy maker proceeds through a succession of incremental changes, he or she avoids serious lasting mistakes in several ways. In the *first* place, past sequences of policy steps provide knowledge about the probable consequences of further similar steps. *Second,* big jumps toward goals are not needed, jumps that would require predictions beyond his or anyone else's knowledge. The policy maker never expects policy to be a final resolution of a problem. Each decision is only one step, one that if successful can quickly be followed by another. *Third,* the policy maker is able to test previous predictions about moves on each further step. *Fourth,* the incrementalist often can remedy a past error more quickly than if policy proceeded through more distinct steps widely spaced in time.

Because practitioners of this approach expect to achieve their goals only partially, they expect to repeat endlessly the sequence just described, as conditions and aspirations change and as accuracy of prediction improves. Any significant problem for planners or policy makers involves tradeoffs among conflicting values, and when dealing with incrementally different alternatives, this is always a problem of adjustments at a margin. But there is no practicable way to state marginal objectives or values except in terms of particular policies. The only practicable way to disclose one's relevant marginal values, even to oneself, is to describe the policy chosen to achieve them. Consequently no overall goal is developed and means and ends are simultaneously chosen. Ideally, rational-comprehensive analysis leaves out nothing important. But it is impossible to take everything important into consideration unless "important" is so narrowly defined that analysis is in fact quite limited. Limits on human intellectual capacities and on available information set definite limits to people's capacity to be comprehensive. Therefore, no one can practice the rational-comprehensive method for really complex problems, and every administrator faced with a sufficiently complex problem must find ways to simplify drastically. In the method of successive limited comparison, simplification is systematically achieved in two principal ways. First, it is achieved through limitation of policy comparisons to those policies that differ in relatively small degree from policies presently in effect. Such a limitation immediately reduces the number of alternatives to be investigated and also drastically simplifies the character of the investigation of each. The second method of simplification of analysis is the practice of ignoring important possible consequences of possible policies, as well as the values attached to the neglected consequences. Almost every interest has its watchdog. Without claiming that every interest has a sufficiently powerful watch dog, it can be argued that the current system often can assure a more comprehensive regard for the values of the whole society than any attempt at intellectual comprehensiveness.

Index (Index Number) A ratio or composite numerical value derived from a series of observations which is used as an indicator or measure of conditions, properties, phenomena or trends. Also, a number used to indicate change in magnitude (as of cost, price or volume of production) of some complex variable as compared with its magnitude at some specified time — usually taken as having a standard of 100.

Index, Economic A measurable aspect of society which indicates the extent to which certain more complex aspects are present. [e.g., the "consumer price index"] Discloses the relative change, if any, of prices, costs, or some similar phenomena between one period of time and some other period of time selected as the "base period." The latter period is usually assigned the index number of 100. There are numerous methods of calculating an index number.

Indicative Planning A method of planning (notably used in France) which primarily restricts itself to analyzing the economy and then making this information available to industry. The modern economy requires the employment of large investments the profitability of which may be profoundly affected by changes in the environment. These changes not only occur rapidly, they are difficult to predict. The individual business has no valid basis for such calculation. The entire economy is interdependent and beyond the analysis capabilities of a single firm. The guiding principle of "indicative planning" is to integrate all the effects of interdependence by, on a nationwide scale, studying the markets and making detailed analyses and forecasts. The instrument used in making these forecasts is the economic input-output analysis. This information is then distributed to the private sector, which is primarily left alone. With this complete information it is assumed that industry will adjust its production, use of resources, and investments to satisfy the projected demands, thus avoiding bottlenecks, smoothing business cycles, and increasing harmony and growth. The planning is implemented by "indicating" to all individuals involved the future consequences of their actions, helping individuals to arrive at a common view on the future of a particular economic activity in relation to the national development aims, demonstrating the advantages of confor-ming to this common view, and only occasionally coercing conformance. Thus, "indicative planning" is not planning for land managers, but planning for managers of societies and economics.

Indicator A measure or measurement of an aspect of a sustainability criterion. A quantitative or qualitative variable that can be measured or described and that, when observed periodically, demonstrates trends. Indicators are quantifiable performance measures used to assess progress toward desired conditions.

Indicator Species (Index Species, Indicator Organism) A plant or animal in a certain location or situation, the presence or absence of which, or its frequency or vigour, is a fairly certain sign that particular environmental conditions are also present, e.g., Escherichia coli bacteria in water indicates probable pollution by human faecal matter. A biologic indicator can be used to refer to signs or symptoms associated with any level of organic organization while indicator species only refers to those associated with the species level. For example, a particular plant might indicate a soil type or the presence or absence of an air or water pollutant. The arrangement of habitats (by tree species and age group) reflects requirements for selected wildlife species.

Indicator, Biologic (Ecological Indicator) An organism or a class of organisms (such as a species) or an ecological community that is so strictly associated with particular environmental conditions that its presence is a fairly certain sign or symptom of the existence of these conditions. In select cases, an organism that exhibits identifiable responses to a pollutant at low levels.

Indifference Curve or Map A two-dimensional graph of a combination of indifference curves denoting an individual's preference system to various combinations of two goods or services. Each line of the family of curves represents equally desirable mixtures of the quantities in question. With different curves representing differing levels of satisfaction it is possible to determine what quantities an individual will consume when faced with different prices and incomes.

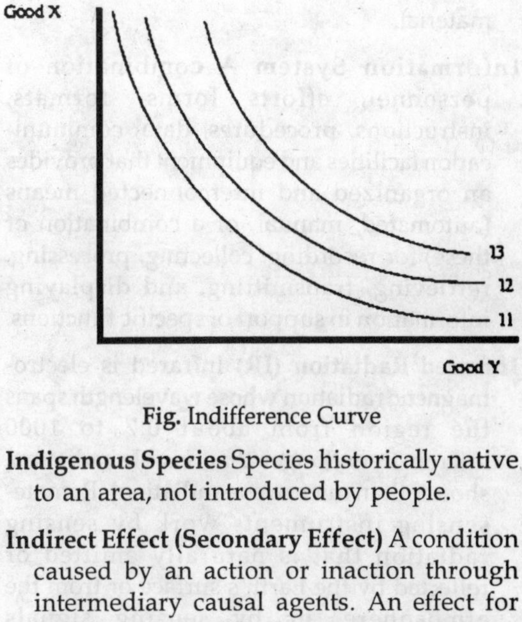

Fig. Indifference Curve

Indigenous Species Species historically native to an area, not introduced by people.

Indirect Effect (Secondary Effect) A condition caused by an action or inaction through intermediary causal agents. An effect for which the causal linkages to the action or inaction are not readily apparent. Contrasts with direct effect. This is not a measure of importance, but merely a classification by causal linkages. Direct effects are usually easier to detect and measure with certainty than indirect effects, but they may be either more or less important than indirect effects.

Individual Distance (Personal Distance) The normal spacing maintained by non-contact animals; the distance two members of the same species experience stress of crowding.

"Individual distance" and personal space interact to affect the distribution of persons. The violation of "individual distance" is the violation of society's expectations.

Infiltration Rate (Infiltration Capacity) The maximum rate at which the soil under various specified conditions (including the presence of an already excess amount of water) can absorb falling rain or melting snow.

Infiltration The flow or movement of water through the soil surface into the ground. This is distinct from percolation which is movement of water through soil layers or material.

Information System A combination of personnel, efforts forms, formats, instructions, procedures, data, communication facilities and equipment that provides an organized and interconnected means (automated, manual, or a combination of these) for recording, collecting, processing, retrieving, transmitting, and displaying information in support of specific functions.

Infrared Radiation (IR) Infrared is electromagnetic radiation whose wavelength spans the region from about 0.7 to 1000 micrometers (longer than visible radiation, shorter than microwave radiation). Remote-sensing instruments work by sensing radiation that is naturally emitted or reflected by the Earth's surface or from the atmosphere, or by sensing signals transmitted from a satellite and reflected back to it. In the visible and near-infrared regions, surface chemical composition, vegetation cover, and biological properties of surface matter can be measured. In the mid-infrared region, geological formations can be detected due to the absorption properties related to the structure of silicates. In the far infrared, emissions from the Earth's atmosphere and surface offer information about atmospheric and surface temperatures and water vapour and other trace constituents in the atmosphere. Since IR data are based on temperatures rather than visible radiation, the data may be obtained day or night.

Infrastructure (Social Overhead Capital) The basic equipment and facilities underlying a nation's, region's, or community's economy (transportation and communications systems, power facilities, schools, hospitals, etc.).

Inherent Erodibility The natural qualities or properties of a soil to erode when subjected to erosive forces.

Initial Action The prompt, pre-planned response to a wildfire.

Input 1. Data that are read into the computer.

2. Land, labor, or capital required for production processes.

3. Anything (often basic resources) that is taken in by or enters into the workings of a process or system, e.g., as nutrients or energy into an ecosystem, effort or information in to a planning process, silt into a stream.

Input-output Analysis (Inter-industry Analysis, Leontief Analysis) A systematic technique for quantitatively analyzing the interdependence of producing and consuming units (sectors) in an economy. It studies the interrelations among producers as buyers of each other's outputs, as users of scarce resources and as sellers to final consumers. The technique has been useful for the study of the way in which the component parts of an economy fit together and influence each other, for short-run forecasting, and for policy guidance. A quantitative study of the interdependence of a group of activities based on the relationship between in puts and outputs of the activities. The basic tool of analysis is a square input output table, interaction model, for a given period that shows simultaneously for each activity the value of inputs and outputs, as well as the value of transactions within each activity itself. It has been applied to the economy and the industries into which the economy can be divided.

INSAT Indian National Satellite.

Insecticide An agent or chemicals, typically used to destroy, prevent, or control insects and similar animals.

Insectivore An animal that eats insects.

Insolation Solar radiation incident upon a unit horizontal surface on or above the Earth's surface.

Instar A stage in the development of an insect between two successive molts.

Institutional Land Uses All buildings, grounds and parking lots composing the facilities for educational, religious, health, correctional and military installations.

Instream Flow The minimum flow necessary for all the uses of water while it is flowing through streams. Some of those uses are fisheries, channel stability, maintenance, riparian habitat maintenance, and aesthetics.

Intangible Value (Imponderable Value) Resource yields which are not directly quantifiable, or if quantifiable, cannot be valued by market mechanisms. The net yields from resource use extend beyond the concept of secondary benefits, and include psychic and indirect monetary benefits to the users. Normally intangible values accrue from the aesthetic, scientific, educational, historical, or recreational aspects of the natural environment.

Integrated circuit (IC) A solid state electronic circuit that consists of several micro-components constructed to perform a special function.

Integrated Pest Management (IPM) The comprehensive systems approach to achieving economical pest damage control in an environmentally acceptable manner. The individual components of integrated pest management in forestry include cultural, mechanical, manual, prescribed fire, biological, chemical, and regulatory means. A strategy which manipulates crop and forest pests to achieve resource management objectives. It is the planned and systematic use of detection, evaluation, and monitoring techniques, and all appropriate silvicultural, biological, chemical, genetic, and mechanical tactics needed to prevent or reduce pest-caused damage and losses to levels that are economically, environmentally, and aesthetically acceptable. Emphasis is on damage management, not necessarily reducing pest population abundance. *Vertebrate damage management* is recommended by the group of people working with those animals and associated problems.

Integrated Resource Management A management strategy that emphasizes no resource element to the exclusion or violation of the minimum legal standards of others.

Interaction Matrix (Matrix) A table with actions, conditions, resources, effects, etc., listed along two perpendicular axes and a matrix structure so created where each position in the matrix corresponds to one element from each axis. By assigning values to the positions in the matrix structure, relations between pairs of elements (one from each axis) can be indicated.

Interdisciplinary Approach The utilization of a team of individuals representing two or more areas of knowledge, learning, or skill working together on the same subject. This approach provides for integrating two or more mental processes into a common, problem-solving effort that is likely to be more effective than the independent thought processes of a single person trained or skilled in one area of knowledge. Contrast with Multidisciplinary.

Interest The cost or time value of money.

Interface A shared boundary. Also a hardware component to link two devices. Also a portion of storage or registers accessed by two or more computer programs.

Interflow That portion of rainfall that infiltrates into the soil and moves laterally through the upper soil horizons until intercepted by a stream channel or until it returns to the surface at some point of infiltration.

Intermediate Cut Any removal of trees from a stand between the time of its formation and the harvest cutting; generally includes cleaning, thinning, release liberation, improvement, sanitation, salvage, and sanitation cuttings. An example is a removal cut which removes seed source trees that are prominent in a seedtree or shelterwood cut. Thinnings are intermediate cuts in an immature stand to control tree density and to improve the growth, quality, yield and species composition of a stand. They may produce income but in some cases add costs.

Intermediate Harvest Any removal of trees from an even-aged stand between the time of its formation and the regeneration cutting.

Intermittent Grazing Grazing range or artificial pasture for indefinite periods at indefinite intervals.

Intermittent Road A road developed and operated for periodic service and closed for more than 1 year between periods of use.

Intermittent Stream (Ephemeral Stream) A stream or portion of a stream that flows only in direct response to precipitation. It receives little or no water from springs and no long-continued supply from melting snow or other sources. It is dry for a large part of the year, ordinarily more than 3 months. The groundwater table lies above the bed of the stream during the wet season but drops below the bed during dry seasons. During wet seasons the stream may receive a contribution from ground water. In many arid areas, the stream channels are always above the water table and therefore carry only surface runoff. Flow may also be interrupted by freezing of the groundwater to some depth below the streambed in smaller streams. A stream that flows only at certain times of the year when it receives water from springs or from some surface sources and is shown by a dashed line on minute quadrangle maps.

Internal Rate of Return The rate of interest at which future project costs arid benefits, discounted to the present, are equal.

International date line An imaginary line of longitude 180o east or west of the prime meridian.

International designator An internationally agreed-upon naming convention for satellites. The designator contains the last two digits of the launch year; the launch number of the year; and the part of the launch, i.e., "A" indicates payload, "B" the rocket booster; or second payload, etc.

International Geophysical Year (IGY) (1957-58) The IGY was organized by the scientific community through the International Council of Scientific Unions (ICSU) . It was highlighted by international cooperation in the exploration of world-wide geophysical phenomena and by the inauguration of the space age through the launching of the first satellites (USSR's Sputnik I and US Explorer 1) to study the upper atmosphere and Earth's nearby environment.

International Space Year (ISY) (1992) Designated the first international celebration of humanity's future in the space age. Themes included the global perspective of the space age, discovery exploration, and scientific inquiry. An important ISY scientific focus was Mission to Planet Earth. A wide range of educational programs and public events emphasized ISY's global perspective.

1992 also commemorated the 500th anniversary of Columbus' voyage to the New World and the 35th anniversary of the International Geophysical Year.

Interplant To set young trees among existing forest growth of similar age and size, planted or natural, to bring the stand to a fully stocked condition.

Interpretive Sites A developed site at which a broad range of natural or cultural history is interpreted or described for the education or enjoyment of the public.

Interspersion Intermixing of cover types, land uses, and conditions for wildlife for meeting year around needs of specific wild animal species. Many different small-size mappable units and many corners (where 3 or more types touch) are indexes to interspersion, an expression of landscape diversity. Only some species respond favourably to interspersion (not all "wildlife").

Intolerant Species 1. Plant species that do not grow well in shade;.

2. Plant and animal species that are unable to endure significant habitat or climate change.

Intrusive Rocks formed from solidification of fluid flowing into or between other rocks. Solidification occurs before surface contacts.

Invader Plant species that were absent in undisturbed portions of the original vegetation and will invade under disturbance or continued overuse.

Invasive Species (Plants) A species not native to the ecosystem under consideration (arriving after major settlement in the US) and if it causes or is likely to cause economic or environmental harm or impair human health.

Inventoried Roadless Areas Those areas identified in a set of inventoried roadless areamaps, contained in Forest Service Roadless Area Conservation, Final Environmental Impact Statement, Volume 2, dated November 2000, which are held at the National headquarters of the Forest Service, or any update, correction, or revision of those maps.

Inventory Gathering of data and processing it on the quantity of physical entities (such as trees, lakes, etc.) in an area.

Inventory, Species A sample survey to estimate the number of flora and/or fauna inhabiting a defined area. The level of resolution of such a listing varies with objectives and may range from only a few conspicuous or predominant species to a complete list for the area. A species inventory does not necessarily constitute an ecological baseline study and the two phrases should not be used interchangeably.

Inversion A stable layer of air having the temperature of the air increase with height.

Ion Atom or molecule that has acquired an electric charge by the loss/gain of one or more electrons.

IPS Inches per second.

Isobars Lines drawn on a weather map joining places of equal barometric pressure.

Isochrons Travel-time zones, concentric lines each of which joins distances with equal travel times from some central reference point of concern.

Isohyetal Line A line drawn on a map joining points that receive the same amount of precipitation.

Isopleth Any lines on a map joining points of equal value, e.g., contour lines, isohyetal lines.

Isothermal Of or indicating equality of temperature.

Isotherms Lines connecting points of equal temperature on a weather map.

Issue A controversial or debatable subject or topic. A subject or question of widespread public discussion or interest regarding management of wildlands. A point in question of law or fact. Something entailing alternatives between which to choose or decide. When the point of controversy can only be vaguely defined or when the causality relationships between use actions and undesirable results are more speculative than documentable, the dispute is referred to as a concern rather than an issue. It may reflect a great difference between an actual (B) and a desired condition (B*) or (B-B*). Within the National Forest System, issues may be 1. A potential factor for determining need for change for a plan;

2. Specific resource concerns related to a proposed action; 3. points of contention or disagreement, or 4. a subject or question of widespread public interest relating to management of the National Forest System.

Issues and Concerns A collective phrase used to include management concerns and public issues relating to management of the wildlands.

Isthmus Narrow strip of land located between two bodies of water; connecting two larger land areas.

Item Type A designation of the way a component should be stored by the computer — whether as alphanumeric characters or as numbers.

ITOS (Improved TIROS Operational Satellite) Second generation, polar-orbiting, environmental satellites utilized to augment NOAAs world-wide weather observation capabilities. ITOS were launched from 1970-1976, but eventually replaced by the third generation of polar-orbiting, environmental satellites TIROS-N (first launched in 1978).

J

Jackson Turbidity Unit (JTU) Standard units for measurement of turbidity based on diffusion of light passed through a water sample.

Jet stream Rivers of high-speed air in the atmosphere. Jet streams form along the boundaries of global air masses where there is a significant difference in atmospheric temperature. The jet streams may be several hundred miles across and 1-2 miles deep at an altitude of 8-12 miles. They generally move west to east, and are strongest in the winter with core wind speeds as high as 250 mph. Changes in the jet stream indicate changes in the motion of the atmosphere and weather.

Joint Education Initiative (JEI) The JEI project was developed by USGS, NOAA, NASA, industry, and teachers to enable teachers and students to explore the massive quantities of Earth science data published by the U.S. Government on CD ROM. JEI encourages a research and analysis approach to science education.

J-Root Seedling or sapling root planted in a manner that forms a J or L configuration in the planting slit. Such plants may grow poorly and die.

Just Compensation The value of property taken for public use, a comprehensive term that includes all elements; but does not exceed market value. It is the value of property at the time of taking plus compensation for delay in payment, if appropriate, plus damages to the owner for value of use of property from date of taking possession to date of judgment if possession is taken by condemnation prior to judgment.

Juxtaposition Principle Often associated with interspersion of wildlife habitat units and the edges of such units, this is the expression of the relative value to a species of two types being adjacent. It expresses the difference to a local population of a named species of wild animals , for example, of a corn field being adjacent to a wheat field vs. a cornfield being adjacent to a fallow field.

K

Key Area Areas of land that supplement or provide major quantities of specific habitat requirements (food, water or cover). For range management usage see Key Grazing Area. For wildlife, these are areas that may limit population growth; have toxic characteristic; be easily eroded; or need intensive care.

Key Components Areas or landscape features particularly important for maintaining the overall integrity of wildlife habitat.

Key Grazing Area That portion of a pasture or grazing unit which, because of its nature, location, and grazing use, serves to control the pattern of grazing use for the pasture as a whole.

Keystone Complex A web of animals and food, predators and prey, that is especially important in defining an ecosystem.

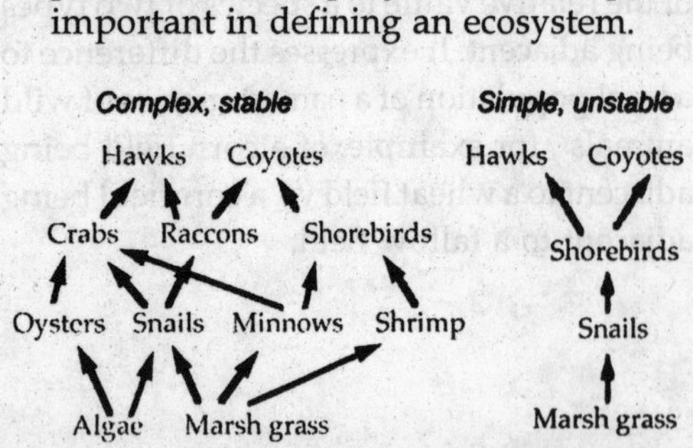

Fig. Keystone Complex

Keystone Species Plants that significantly modify the physical environment or animals that have important influences on other biota; e.g., removal of a keystone tree species from swamp islands may result in loss of an animal species there and the entire community.

Kilohertz (kHz) One thousand hertz, i.e., one thousand cycles per second.

Kilometre (km) Metric unit of distance equal to 3,280.8 feet or .621 statute miles.

Kilopascal (or megapascals), a unit of expression for osmotic concentrations (e.g., of urine and blood), typically with negative values.

Knot Unit of speed of one nautical mile (6,076.1 feet) an hour.

Knutson-Vandenberg Deposits Payment (according to a bill by K-V) by a timber purchases included in high bid price that is retained by the Forest Service for reforestation and other forest improvements.

L

Lacustrine Includes wetlands and deepwater habitats with the following characteristics: 1. Situated in a topographic depression or a dammed river channel;.

2. Lacking trees, shrubs, and persistent emergents with greater than 30 percent aerial coverage; and.

3. Total area exceeds 20 acres. Similar wetland and deepwater habitats < 20 acres in size are included where water depth exceeds 6.6 feet at low water. The typical lacustrine system of a forest includes permanently flooded lakes and reservoirs formed by damming a river channel. It is bounded by a contour approximating the normal spillway or pool elevation.

Lake A body of fresh or salt water entirely surrounded by land. An inland body of standing water, an expanded part of a river, or an impoundment formed by a dam. "Small lakes" are fishing impoundments, usually 5 to 25 acres (2 - 10 hectares) in area.

Land A term denoting the entire complex of surface and near-surface attributes of the solid portions of the surface of the Earth which are significant to people. Water bodies occurring within landmasses are included with land. A specific area of the Earth's surface. Its characteristics embracing all reasonably stable, or predictably cyclic attributes of the biosphere directly above and below this area including those of the atmosphere and climate, the soil and underlying geology, the topography, the hydrology, the plant and animal populations, and the results of past and present human activity, to the extent that these attributes exert a significant influence on present and future uses of the land by people. It is all but the purely socioeconomic (human) attributes of the environment. It is assumed that all approaches to interpretative land classification would, to a varying extent, take additional socioeconomic factors into account but these are not considered to be attributes of the "land" itself. "Land" is a broader concept than soil. Within economic theory it is one of the major factors of production consisting of a good that is supplied by nature without the aid of people. The term may include not only the Earth's surface, both solid land and water, but also anything that is attached to the Earth's surface. Thus, all natural resources in their original state, such as mineral deposits, wildlife, timber, and fish are land; so are sources of energy, outside of people themselves, such as water, coal deposits, and the natural fertility of the soil.

Land and Water System One of the six "systems" established by the Forest Service to have a systematic, orderly way to view and evaluate its many diverse but interrelated activities. The role of this system is to protect, conserve, and enhance the basic resources of air, soil, and water on forest and rangeland. It provides the base on which the other resources grow. Activities on National Forest land include land-use planning, administration of special land uses, easements, and land adjustments, and management of mineral areas in addition to the protection, conservation, and enhancement role. The six "systems" are "Land and Water", Timber Resource, Outdoor Recreation and Wilderness, Rangeland Grazing, Wildlife and Fish Habitat, and Human Community Development.

Land breeze A nocturnal coastal breeze that blows from land to sea. In the evening the water may be warmer than the land, causing pressure differences. The land breeze is the flow of air from land to sea equalizing these pressure differences.

Land Capability The inherent ability of land to be used without permanent damage. Land capability, as ordinarily used is an expression of the effect of physical land conditions, including climate, on the total ability to be used without damage for crops that require regular tillage, for grazing, for woodland, and for wildlife. Land capability involves considering 1. The risks of land damage from erosion and other causes and. 2. The difficulties in land use owing to physical land characteristics, including climate.

Land Capability Class One of the eight classes of land in the land capability classification system of the Soil Conservation Service, now the Natural Resource Conservation Service. The eight land capability classes in this system are distinguished according to the risk of land damage or the difficulty of land use. The first four classes distinguish land inherently able to be used for cultivation and other uses. Soils in Class I have few limitations that restrict their use. Soils in Class II have some limitations that reduce the choice of plants or require moderate conservation practices. Soils in Class III have severe limitations that reduce the choice of plants or require special conservation practices, or both. Soils in Class IV have very severe limitations that restrict the choice of plants, require very careful management, or both. The second four classes distinguish land generally inherently unable to be used for cultivation (without major treatment). Soils in Class V have little or no erosion hazard but have other limitations, impractical to remove, that limit their use largely to pasture, range, woodland, or wildlife food and cover. Soils in Class VI have severe limitations that make them generally not able to be used for cultivation and limit their use largely to pasture or range, woodland, or wildlife food and cover. Soils in class VII have very severe limitations that make them not able to be used for cultivation and that restrict their use largely to grazing, woodland, or wildlife. (VIII). Soils and landforms in class VIII have limitations that preclude their use for commercial plant production and restrict their use to recreation, wildlife, water supply, or aesthetic purposes.

Land Capability Classification A grouping of kinds of soil into special units, subclasses, and classes according to their capability for intensive use and the treatments required for sustained use.

Land Capability Map A map resulting from capability classification showing land capability units, land capability subclasses, and classes or a soil survey map.

Land Capability Subclass Groups of land capability units within classes of the land capability classification system that have the same kinds of dominant degree and kind of limitations for agricultural use as a result of soil and climate, e.g., some soils are subject to erosion if they are not protected, while others are naturally wet and must be drained if crops are to be grown. Some soils are shallow or droughty or have other soil deficiencies. Still other soils occur in areas where climate limits their use. The four kinds of limitations recognized at the subclass level are risks of erosion; wetness, drainage, or over flow; other root zone limitations; and climatic limitations.

Land Capability Unit A group of soils that are nearly alike in inherent ability to be used for plant growth and in their responses to soil management. Capability units provide more specific and detailed information application to specific fields on a farm or ranch than the "subclass" of the land within a capability classification.

Land Classification Land classification is the process whereby the complex of surface and near-surface attributes of the solid portions of the Earth's surface are identified and organized into some system of mappable units according to a set of criteria or principles of relatedness. "Land classification" creates a framework of generalization about the complexity of "land" properties which enables common characters (primarily for named uses) to be defined and described and units with similar properties to be regarded as equals although geographically separate. Classification, before computers, was a prerequisite to any transfer of knowledge about the relationships between land properties and land use suitability and capability from one unit to others, i.e., if there was to be predictability in land use planning. Computer mapping allows the chara-cteristics of each small land unit to be analyzed, usually reducing the need for or utility of prior classification.

Land District A relatively homogeneous unit of land characterized by a distinctive pattern of relief, geology, geomorphology, and regional vegetation. This is the second largest of the four classification levels of the Canadian Biopysical Land Classification system. The largest unit is the land region and the two smaller units in order of decreasing size are land system and land type.

Land Evaluation (Terrain Evaluation) The assessment of the inherent capability and managed suitability of land for people's use in agriculture, forestry, engineering, hydrology, regional planning, recreation, etc.

Land Form Sometimes used when referring only to the three dimensional shape of the ground surface – without any reference to the processes responsible for that shape. "Land form" may have a different meaning than landform when used in geomorphological contexts. "Land form" is always used for surface features, the origin of which can be attributed to particular geological processes or structures. A type of land surface which exist as a result of geological activity, such as a plateau, plain, basin, or mountain.

Land Management The intentional process of planning, organizing, programming, coordinating, directing, controlling, and adapting land use actions to achieve stated objectives.

Land Region An area of land with a certain type of vegetation that manifests a distinctive climate. This is the largest unit of the four classes of the Canada Biophysical Land Classification System. The other three classes in order of decreasing size are land district, land system, and land type.

Land Resource Area Broad geographic areas having similar soil, climatic, geologic, vegetative and topographic features, and intensities of problems in soil and water conservation (larger than a land resource unit and smaller than a land resource region).

Land Resource Region A generalized grouping of land resource areas reflecting regional relationships to agriculture with emphasis on soil and water conservation.

Land Resource Unit A subdivision of a land resource area with emphasis on a specialized type of agriculture, intensities, or problems in soil and water conservation. It has a narrower range in relationship to agriculture (than a land resource area) with emphasis on soil and water conservation.

Land System 1. An area of land with a recurring pattern of landform having homogeneous soil characteristics and supporting a particular type of vegetation.

2. The second smallest of the four classification levels of the Canada Biophysical Land Classification System.

Land Systems Inventory A land classification system based on geomorphic principles. It is based on an understanding of geologic structure, land form genesis and geomorphic processes (as reflected in land surface form and features), individual kinds of soil and the factors which determine the behaviour of ecosystems (i.e., climate, vegetation and animal life, relief, parent materials and time). The hierarchical levels in the "land systems inventory" classification scheme are: 1. the "physiographic province" (a division of the land in which the pattern of topographic elements of altitude, relief and type of land forms are uniform throughout an area of several thousand square miles or larger);.

2. The "section" (land units in which the combined factors of geologic structure, geomorphic processes, climate and time have been essentially the same and thus produced a uniform pattern of topography over areas of hundreds to several thousand square miles);.

3. The "subsection" (the smallest land unit which can be identified using the basic criteria of uniform geologic factors and climate operating over time to delineate essentially uniform areas of tens to hundreds of square miles);.

4. The "land type association" (units tens of square miles in size over which the uniformity of geomorphic processes has produced different types of landscapes which can be visually identified on maps, air photos or on the ground);.

5. The "landtype" (visually identifiable unit areas of a tenth to one square mile resulting from homogeneous geomorphic and climatic processes which have produced defined patterns of soils and vegetation potentials);.

6. The "landtype phase" (0.01 to 0.1 square mile areas delineated on the basis of being characterized by uniform, specific kinds of soils or soil phases or sometimes on the basis of physiographic, vegetative or bedrock characteristics) and;.

7. The "site" (areas of less than an acre defined and characterized by the occurrence of individual kinds of soils or discrete plant communities.).

Land Tenure The holding of land and the rights that go with such holding, including all forms of holding from fee simple title (embracing all possible rights within the general limitations imposed by the government) to the various forms of tenancy of group ownership with allocation procedures, various kinds of lending possibilities, etc.

Land Type An area of land Canada Biophysical Land Classification System having a fairly homogeneous combination of soil features at a level corresponding to the soil series, i.e., a group of soils having horizons similar in distinguishing characteristics and arrangement in the profile and developed from the same parent material and chronological sequence of vegetation. The land type is the basic biophysical land unit that can be interpreted to provide land evaluation data useful for planning and development. The mapping scale at the land type level ranges from 1/20,000 to 1/10,000. The dominant ecological variables are soil moisture regime, physical and chemical soil properties, and micro-topography. The related units in order of decreasing size are land region, land district, land system, and land type.

Land Units A portion of a landscape unit, i.e., physiographic site types and physiographic site phases having been grouped into 16-sq. mi. minimum units, with special significance for a specific use.

Land Use The occupation or reservation of land or inland water for any human activity or any defined purpose. It also includes use of the air space above the land or water.

Land Use Allocation Committing a given area of land or a resource to one or more specific uses, e.g., to campgrounds, wilderness areas.

Land Use Constraints Any factor that may act to reduce management options or discourage various uses, e.g., erosive soils, unstable slopes, steep topography, presence of archaeological or historic remains.

Land Use Plan A coordinated composite of information, ideas, policies, programs, and activities related to existing and potential distribution and uses of land within a given area, and stating how objectives will be achieved. It identifies where activities take place or where additional activities could take place under future conditions or circumstances. It provides the basis for the complete coordination of existing and future programs and activities with adjoining lands.

Land Use Planning To create land use plans. It is the continual process of organizing the development and use of lands and their resources in a manner that will best meet the needs of people over time while maintaining maximum flexibility for a dynamic combination of desired resource outputs for the future. It includes inventorying and assessing the status, potentials, and limitations of a particular area and its resources; interacting with the populations associated and/or concerned with the area to determine their needs, wants, and aspirations for the future; hypothesizing patterns of use compatible with both the area's capabilities and human needs and wants; proposing and assessing the consequences of, and implementing the best of alternatives for achieving the desired patterns of use, documenting the history of use to enable comparison of expected and experienced consequences; and repeating the entire process making adaptations for changing environmental status, technology, and human wants.

Land Use Practice (Discrete Land Use Operation) The various kinds of altering actions and/or activities associated with or utilized in facilitating or making possible different types of land uses, e.g., clearing of vegetation, land grading, paving, landscaping, fencing, fertilizing, irrigating, etc. "Land use practices" are distinct from land use types. "Land use types" is a classification system based on the purpose for which the land is being used, e.g., timber production, range, second home subdivisions, etc. "Land use types" are bundles of more or less shared "land use practices" — though the precise types and amounts of practices that are utilized in conjunction with a specific use type may vary from one specific instance to another. "Land use practices" is an alternative to "land use types" as a way of classifying for study and/or regulation the problems created by use. By restricting practice alternatives the problems associated with specific uses can sometimes be eliminated or reduced while not necessarily preventing use— because there may be an alternative acceptable practice.

Land Use Survey A survey of the uses to which land is put in a particular area, usually summarized both in map form and statistically.

Land Use Type (Land Use) The primary use of a tract of land, e.g., crops, pasture, forest, urban, and other.

Land, Marginal Land of questionable abilities to be used economically for a specific purpose.

Land, Submarginal Land that does not return enough to pay costs of ownership or operation for a specific use.

Land,Alienated Lands of one ownership enclosed within the boundaries of another ownership. Often refers to land in private ownership within the boundaries of public land.

Landline A property boundary.

Lands Suitable for Grazing or Browsing Lands with vegetation that can be used by domestic and wild herbivores without damage to the soil and water resource values.

LANDSAT Land Remote-Sensing Satellite, operated by the U.S. Earth Observation Satellite Company (EOSAT). Commercialized under the Land Remote-Sensing Commercialization Act of 1984, LANDSAT is a series of satellites (formerly called ERTS) designed to gather data on the Earth's resources in a regular and systematic manner. Objectives of the mission are: land use inventory, geological/mineralogical exploration, crop and forestry assessment, and cartography. LANDSAT has a spatial resolution of 28.5 meters.

Restructured Federal agency responsibilities for the LANDSAT program are effective for the acquisition and operation of LANDSAT 7. New operating policy specifies that NOAA will be responsible for satellites after they are placed in orbit, NASA will be responsible for the development and launch of LANDSAT 7, and that the U.S. government will provide unenhanced data to users at no cost beyond the cost of fulfilling their data request.

Landscape A total part of Earth (or other planets), a composite of all of the characteristics that distinguish a certain area on the Earth's surface from other areas. An expanse of Earth typically seen within one viewing, but often a large designated area on a map somehow different than contiguous areas.

Landscape Ecology The study of spatial and temporal variety (heterogeneity) in the structure, dynamics, and relations of plants, animals, (including people), and landscape elements at a large scale.

Landscape Management Planning, administering, and controlling the use and development of land so that the visual effects maintain or upgrade people's psychological welfare.

Landscape Planning A portion of the land use planning process which deals with large-area physical (i.e., biological, geological and hydrologic) values, aesthetic and cultural (i.e., historical and anthropological) values, and with the relationships among these values and land uses.

Landscape Unit A grouping of physiographic site types and physiographic site phases into large (16 sq. mi. minimum) units to facilitate making large-scale planning decisions.

Landscape, Canopied The type of visual landscape to be found under an extensive, continuous crown cover of trees – as within a forest. The overhead foliage and branches define the ceiling plane and the perspective closure of tree stems or screens of understory shrubs and young trees or landforms define the side enclosures. This is one of the six general types in Litton's (1968) scenic composition classification system. The others are panoramic landscape, feature-dominated landscape, enclosing landscape, focal landscape, and detailed landscape.

Landscape, Cultural The natural landscape modified by human occupation.

Landscape, Detailed The close view of objects or features, particularly surface characteristics, colour, or touchable nature. There is little or no sense of space and attention is restricted to the intimate details of features.

Landscape, Enclosing A small outdoor space (rather than strictly a view type) with definite visual sidewalls or edge lines (formed by terrain, tree, lake or meadow surfaces) which enclose or define the field of vision. The viewer has a sense of presence by being within the foreground that is below the tops of the sidewalls.

Landscape, Feature-dominated A view in which a single feature (or landmark), either natural or made by people, dominates the landscape because of sharp contrast with the general surroundings.

Landscape, Focal A view characterized by convergent lines which lead the eye to the principal point of their apparent origin.

Landscape, Panoramic An area with an unobstructed or complete view or a continuous series of scenes extending to the distant horizon. The visual feeling is one of expanding or unbounded space, receding directly ahead toward the limits of distance on visibility and also extending laterally to the limits of peripheral vision. Although indicating the availability of a view in every direction, inherent physiological limitations on the extent of what an observer can see without turning the head suggests that a possible view of 180 degrees is a more likely condition for defining a "panoramic landscape". The term is reserved for distant or background scenes which are under observation, not horizontal scene compositions in the middle ground or foreground.

Landslide (Landslip) Downslope movement of a relatively dry or coherent mass of earth and/or rock along an interior surface slope at a rate fast enough to be readily perceived. Movement of a wet mass is called a mudslide.

A LANDSLIDE, WITH CROSS-SECTION

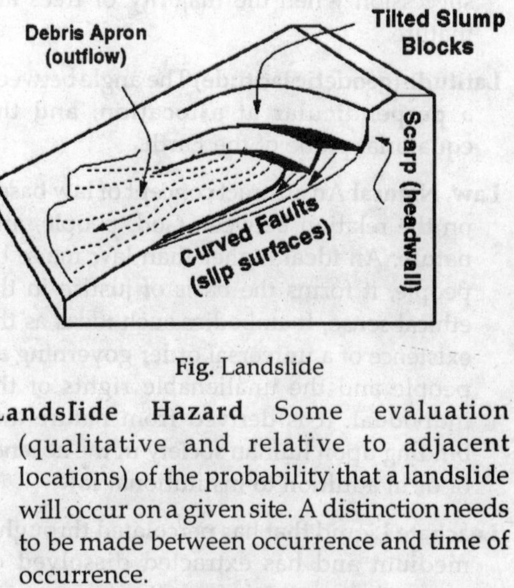

Fig. Landslide

Landslide Hazard Some evaluation (qualitative and relative to adjacent locations) of the probability that a landslide will occur on a given site. A distinction needs to be made between occurrence and time of occurrence.

Landtype (Watertype) Units created by isolating areas of differing landform, geologic composition and water content within the site region. A "landtype" is a subdivision of a "site region" and determined by certain soil properties (such as texture depth, mineral composition, water content, etc.), climate, and the nature of parent material in large-scale land features. Landtype units range in size from about one-tenth to one square mile. Their size and composition depend upon the significance of physical characteristics that can be readily interpreted to identify hazard, capability, and productivity potentials that are reliable for land use planning purposes.

Laser (light amplification by stimulated emission of radiation) Active instrument that produces discretely coherent pulses of light (light waves with no phase differences, or with predictable phases differences, are said to be coherent).

Laser ranging The use of lasers to measure distances.

Late Forest Succession A stage of forest succession when the majority of trees are mature.

Latitude (geodetic latitude) The angle between a perpendicular at a location, and the equatorial plane of the Earth.

Law, Natural An abstract concept of law based on the relation between God, people, and nature. An ideal higher than law made by people, it forms the basis of justice in the ethical sense. It embodies such ideas as the existence of a universal order governing all people and the unalienable rights of the individual. It is derived from nature and binding upon human society in the absence of or in addition to institutional law.

Leachate Liquid that has percolated through a medium and has extracted dissolved or suspended materials from it.

Leaching The removal of materials in solution by the passage of water through a solid medium, e.g., through soil or overburden.

Leasable Minerals Coal, oil, gas, phosphate, sodium, potassium, oil shale, and geothermal steam.

Least Cost The minimum cost that must be accrued to meet specified goals and objectives.

Left-justified A string of characters, such as a title, positioned within a column in such a way that its first character falls in the left-most position of the column.

Legal Institution A society's system of legal rules, principles, and agencies.

Legend A listing that contains symbols and other information about a map.

Lentic Pertaining to standing (nonflowing) waters such as lakes, reservoirs, ponds, and swamps.

Lexicography A method of choosing among alternatives based on an analoguey to the way words are arranged in a dictionary. A single objective or attribute is rated as most important, and alternatives are ranked according to how well they satisfy this objective. If an alternative is clearly superior in this attribute to other alternatives it is selected. Otherwise the second most important attribute is used to rank the top alternatives from the first ranking. This process continues, constantly reducing the number of alternatives ranked and ranking by less important attributes, until a single project is clearly superior for the attribute being used to rank the alternatives.

Life Cycle The successive stages through which an organism passes from the spore or fertilized egg of one generation to the spore or fertilized egg of the next generation. A continuous, descriptive account of a life cycle is called the life history of an organism.

Life Style A characteristically different way of living which may be an individual variant within the cultural main stream or may be an individual expression of a subculture. "Life styles" are generally expressed through the means of economic sustenance, dwelling site and type, types of group associations, and social practices such as family form, religious practices, sexual mores, style of dress, type of diet, etc.

Life Zone An altitudinal, latitudinal, or climatic region or belt with distinctive faunal and floral characteristics, type of animal and plant life found in them – e.g., the Upper Sonoran, Transition, Canadian, Hudsonian, and Alpine life zones.

Light 1. Form of radiant energy that acts upon the retina of the eye, optic nerve, etc., making sight possible. This energy is transmitted at a velocity of about 186,000 miles per second by wavelike or vibrational motion.

2. A form of radiant energy similar to this, but not acting on the normal retina, such as ultraviolet and infrared radiation.

Interplay between light rays and the atmosphere cause us to see the sky as blue, and can result in such phenomena as glows, halos, arcs, flashes, and streamers.

Lightning A discharge of atmospheric electricity accompanied by a vivid flash of light. During thunderstorms, static electricity builds up within the clouds. A positive charge builds in the upper part of the cloud, while a large negative charge builds in the lower portion. When the difference between the positive and negative charges becomes great, the electrical charge jumps from one area to another; creating a lightning bolt. Most lightning bolts strike from one cloud to another; but they also can strike the ground. These bolts occur when positive charges build up on the ground. A negative charge called the "faintly luminous streamer" or "leader" flows from the cloud toward the ground. Then a positively charged leader; called the return stroke, leaves the ground and runs into the cloud. What is seen as a lightning bolt is actually a series of downward-striking leaders and upward-striking return strokes, all taking place in less than a second.

Lightning bolts can heat the air to temperatures hotter than the surface of the sun. This burst of heat makes the air around the bolt expand explosively producing the sound we hear as thunder Since light travels a million times faster than sound, we see lightning bolts before we hear their thunderclaps. By counting the seconds between a flash of lightning and the thunderclap and dividing by five, we can determine the approximate number of miles to the lightning stroke.

Limb viewing (occultation) The process of viewing the atmosphere at a tangent to the Earth's surface. The viewing signal, from a star or another satellite, is occulted or obscured by the intervening atmosphere. The absorption of light from the sun or star provides information on the properties of the atmosphere at different heights. Limb viewing instruments can also sense infrared or microwave-emitted radiation from the atmosphere.

Limiting Factor Any environmental factor whose presence, absence, or abundance is the main factor restricting the distribution, numbers, or condition of an organism. A factor to which the limit of tolerance of an organism is first reached and which, therefore, acts as the immediate restriction to one or more of its functions or activities, thus its abundance and geographic distribution.

Limits The number of livestock that may be grazed under provisions of a grazing preference.

Linear Program Model A deterministic mathematical method (a computer program) used to determine the best use of resources to achieve a desired result and limitations on available resources can be expressed in the form of equations. The restrictions on its attainment are expressible as a system of linear equalities or inequalities (e.g., $y=a+bx$ or $a>b$). There are linear behavioural relationships and an optimal solution is sought (by maximizing or minimizing a system output) subject to limits or constraints.

Litter The uppermost layer of organic debris on the ground under a vegetation cover, i.e., essentially the freshly fallen or only slightly decomposed vegetable material, mainly from foliage but also bole, bark fragments, twigs, flowers, fruits, etc.

Littoral Belonging to the shore of seas; the word applied loosely and generally to the seashore as well as to the shores of lakes and rivers.

Local Analysis A characterization of the ecological, social, and economic components for various times and locations for a smaller area than that of a broad-scale assessment. Local analyses often tier to broad-scale assessments. Local analyses provide comprehensive descriptions of ecological system structure, process, and functions. The geographic area of a local analysis and its data resolution depend on the topics of general interest or concern being addressed. Like broad-scale assessments, local analyses represent a synthesis of current scientific knowledge including a description of uncertainties and assumptions; however, they also provide for the gathering of new information which can be used in developing site-specific projects.

Local Economic Effects Effects of an action or inaction in the resource area on financial transactions in the immediate area of the action or inaction. The "local" area would include an area encompassing the primary sources of employment and retail outlets for basic commodities such as food utilized by the people of that area.

Local Overgrazing Overgrazing in localized areas on range, as, for example, near watering places.

Local Road Roads and trails constructed and maintained for, and frequently by, the single-purpose activities. They connect terminal facilities with forest arterial roads or public highway. Their location and standard are usually determined and controlled by a specific resource activity rather than travel efficiency. Forest local roads may be developed and operated for either long-term or short-term service.

Location (Mineral Location) The acts of the general mining laws by which the right of exclusive possession of locatable mineral deposits is vested in the locator. A "mining claim" may refer to a parcel of land containing soil or rock which has value because of its chemical composition, while "location" is the act of appropriating such land according to certain established rules.

Location Theory The theory which attempts to explain and predict the location of individual, commercial, and industrial development. It is important to the planner for predicting population and industrial distribution that would result from alternative plans.

Log A piece of the woody stem of a tree. The trunk portion of a tree used for sawlogs. Typically 16 feet long. A tree with 1 and a half logs has a 24-foot bole before major limbs or a minimum diameter limit are encountered.

Log Deck Also called log landing, log yard, brow or bunching area. A place where logs or tree-length material is assembled for loading and transporting.

Log Rule A device, usually presented in tabular form, which expresses log volume content based on log diameter (inside bark of the small end) and length. A *log rule* expresses the volume of cut logs. A *tree rule* expresses the volume of standing trees.

Logarithm Exponent of the power to which it is necessary to raise a fixed number (the base) to produce the given number. For example, the logarithm of 100 (base 10) is 2 because 102 = 100.

Logged-over Forest A forest in which most or all of the merchantable timber has been removed.

Logging Cutting and extracting timber, particularly logs and pulpwood. "Harvesting", beyond removing logs, may be expanded to include the cutting, initial processing if any, and extraction of any forest product.

Logging, Tractor Any logging method that uses a tractor as the power for transporting logs from the stumps to a collecting point, whether by dragging or carrying the logs.

Logical Entry All the data about a single major item in a database. (For instance a data collection site.).

Longitude The angular distance from the Greenwich meridian (0°), along the equator This can be measured either east or west to the 180th meridian (180°) or 0° to 360°W.

Long-range Planning Planning for a system for a period from the present of at least 50 years.

Long-Term Action governed by a plan generally taking place over a period (specified) longer than 20 years from the present.

Long-term Effects Those effects that generally occur after 20 years.

Long-Term Sustained Yield Timber Capacity The highest uniform wood yield from lands being managed for timber production that may be sustained under a specified management intensity and set of constraints and objectives.

Lop 1. To chop branches, tops, or small trees after felling so that the slash will lie close to the ground.

2. To cut the limbs from a felled tree.

Lotic Flowing waters, including creeks, streams, and rivers.

Low A logic state corresponding to a binary "0". Satellite imagery is displayed on a computer monitor by a combination of highs and lows.

Low or low-pressure system A horizontal area where the atmospheric pressure is less than it is in adjacent areas. Since air always moves from areas of high pressure to areas of low pressure, air from these adjacent areas of higher pressure will move toward the low pressure area to equalize the pressure. This inflow of air toward the low will be affected by the Earth's rotation and will cause the air to spiral inward in a counterclockwise direction in the northern hemisphere. The air eventually rises near the center of the low, causing cloudiness and precipitation.

The air in a low rotates in a counterclockwise direction in the Northern Hemisphere, and in a clockwise direction in the Southern Hemisphere. Low-pressure cells are called cyclones.

M

Macroclimate The general large-scale climate of a large area or country, as distinguished from the mesoclimate and microclimate.

Macroeconomics Economic studies or statistics that consider aggregates of individuals or groups of commodities; for example, total consumption, employment, or income.

Macrophyte Any plant that can be seen with the naked, unaided eye; e.g., aquatic mosses, ferns, liverworts, rooted plants, etc.

Macrophyton Large aquatic plants that can be seen without magnification, including mosses and seed plants.

Magnetosphere Region surrounding a celestial body where its magnetic field controls the motions of charged particles. The Earth's magnetic field is dipolar in nature. That is, it behaves as if produced by a giant bar magnet located near the center of the planet with its north pole tilted several degrees from Earth's geographic north pole.

The Earth's magnetic field presents an obstacle to the solar wind, as a rock in a running stream of water This obstacle is called a bow shock. The bow shock slows down, heats, and compresses the solar wind, which then flows around the rest of Earth's magnetic field.

Maintenance The personnel and costs required to ensure the continued safe operation and/ or utility of system capital facilities. The word has been used for personnel as well as for the entire system.

Maintenance Level A formally established criterion that prescribes the intensity of maintenance necessary for the planned operation of a road.

Major Ions Elements which are (or could be) in fairly high concentration in most natural waters, such as calcium, magnesium, sodium, potassium, bicarbonate, carbonate, sulphate, and chloride.

Major Land Use A grouping of primary land uses into classes with similar characteristics, e.g., cropland, pasture, and forest.

Major Recreation Facility Developed recreation facilities with substantial investment in facilities or operation costs or receiving significant recreation use. Examples are campgrounds, picnic areas, swimming areas, boating sites, and interpretive sites.

Major Vegetation Types Plant communities, which are typically named after dominant plant species that are characteristic of the macroclimate and geology of the region or subregion.

Majority Determinism A point of view in resource planning which advocates making resource use and allocation decisions on the basis of a simple expression of the most popular choice of use or mixes of uses. This simple one-person-one-vote approach has severe problems in the land-use-planning context on several grounds. How are the alternative resource use mix choices formulated and presented to the "voters"? How is the planned-for populace defined, located and canvassed? How are the desires of transient resource users (such as campers and hikers who do not belong to organized groups) to be counted, and how are they to be weighed against the needs and desires of permanent residents and users?.

Management The action or activities of analyzing, preparing, organizing, and directing environmental and human systems and their relations to achieve specified human objectives; controlling systems toward pre-determined end states.

Management Area 1. An area of land that has common direction and is operated to achieve a common set of objectives throughout.

2. A specifically identified area within the "plan area" to which specific plan components (desired conditions, objectives, identification of suitable and unsuitable land uses, or special designations) are applied.

Management Concern A matter of importance to the management of a land area, an issue or problem needing solution, usually identified internally by an agency or owner.

Management Direction A statement of goals and objectives and the management prescriptions and the associated standards and guidelines for governing them.

Management Emphasis The specific resource outputs or benefits to be featured or enhanced.

Management Information Data that are classified as neither an output nor an activity but are included in reports to allow managers to monitor progress; e.g., motor vehicle accident frequency used in a status report.

Management Intensity The extent of a practice or combination of management practices designed to obtain different levels of goods and services.

Management Objective A concise statement that describes a desired condition or function of the land to be achieved sometime in the future.

Management Opportunity A statement of general actions, measures or treatments that address the public issue or management concern in a favourable way.

Management Practice A specific action, measure or treatment.

Management Prescription Statement of actions to take, and practices selected and scheduled for use in a specific area to attain stated benefits.

Management Program Includes, but is not limited to, a comprehensive statement in words, maps, illustrations, or other media of communication, prepared and adopted setting forth objectives, policies, and standards to guide public and private uses of lands and waters. It is equivalent to a "plan" in some usage. Also the action itself.

Management Situation A section of the planning area guide which contains a comprehensive statement of the planning area resources, its history as it may influence planning, past and present uses, and a review of the publics directly concerned with the area. The significance of the information about the area as it relates to national objectives and program targets is also important as it describes how that area is unique in its ability to contribute to satisfying human needs.

Management Type A forest type that has been selected as the species to manage for timber yields. Normally the species is selected based on the best timber yield but expected present net value may be a more suitable criterion.

Management Unit The finest delineation of land area, which identifies the area of land or water to which specific management decisions apply.

Management Zone For planning purposes, areas of land with similar characteristics, whether they are social, economic, or environmental characteristics, may be delineated into management zones. Such stratification is used for planning areas to help delineate the particular area of land on which certain coordinating criteria apply.

Management-by-Objective A management process whereby managers at various levels in an organization jointly define the organization's goals, define each individual's major responsibilities in terms of measurable results, and use these measures as guides for operating the organization and assessing unit or individual contributions to meeting the identified goals.

ManagementPlan A written plan for the operation of a forest property using forestry principles. It usually reconds data and prescribes measures designed to provide for optimum use of all forest resources.

Map Overlay (Overlay Map) A transparent sheet accompanying a map, on which limited information, colouring, or symbols are entered so that when the sheet is placed on the map, the effect is identical to having entered the overlay information on the map itself. By combining several overlays, it is possible to display a variety of data combinations and classifications that would not be possible otherwise. The map may display physical or computer images. The spatial distribution of data combinations, or more exactly the areas of joint occurrence and dissimilar occurrence, may be seen on physical or computer maps.

Map, Operational A precise, detailed, large-scale map, usually with small contour intervals but sometimes only planimetric (i.e., representing only the horizontal position of features), constructed for specific uses – such as administrative site development, road location, bridge sites, dam sites, campground development, ski areas, and geological area maps.

Map, Planimetric A map representing the horizontal position of features, in contrast to a topographic map that shows both the horizontal and vertical position of features.

Marginal Acres The component of the regulated commercial forest land that includes areas not qualifying as standard or special components primarily because of excessive development costs, low product values, resource protection constraints, and inadequate markets.

Marginal Analysis A type of analysis in which the only costs and benefits considered are those about which decisions can be made. Fixed benefits and costs are not considered.

Marginal Cost The addition to total cost caused by producing one or more unit of output.

Marginal Land Land of questionable economic capabilities for a specific purpose, probably at the financial margin, within one unit of profitability.

Market Place An hypothetical construct of economic theory representing the place all potential buyers and sellers of all goods and services come together, express their desires and offer their wares, through bargaining establish a price structure and production schedule, and complete their transactions. All locations where economic transactions occur treated as a single whole.

Market Value (Market Price) A price at which both buyers and sellers are willing to do business; the market or current price. The price at which the perceived supply and demand are equal.

Market Value Quantitative values for resources in public demand which are traded in dollars, e.g., timber, minerals, and some types of recreation.

Market-Valued Outputs Goods and services valued in terms of what people are willing to pay for them, as evidenced by market transactions (money exchange; these include timber sales and special use permit fees). Non-market-valued outputs are not evidenced by market transactions. Their value has to be estimated by other means. These outputs include dispersed recreation and wilderness use.

Marking Timber The process of indicating what trees are to be cut or otherwise treated. Prior to timber sales it is advisable to mark with paint each tree to be harvested. One spot of paint at eye level and one on the stump portion will help determine whether unmarked trees have been cut.

Marsh A wetland, part of the palustrine system, dominated by emergent, submerged, or floating aquatic vegetation.

Mass Diagram A graphical representation of cumulative quantities (e.g., soil or snow) over time. Each point on the curve is the sum of all preceding quantities.

Mass Instability (Slope Stability) An evaluation of the tendency for the materials on a slope (e.g., rocks, soil, and snow) to move downhill as a large mass.

Mass Movement Downslope unit movement of a portion of the land's surface, i.e., a single landslide or the gradual simultaneous downhill movement of a whole mass of loose earth material on a slope face.

Mass-Wasting A general term for any of the variety of processes by which large masses of earth material are moved downslope by gravitational forces – either slowly or quickly.

Mast *Hard mast* is the hard-shell fruit or nuts of oaks, beech, walnuts, chinquapins, and hickories. *Soft mast* includes the fruit and berries of dogwood, viburnums, elderberry, huckleberry, grape, raspberry and blackberry. Mast connotes that animals consume these.

Master Plan A comprehensive, long-range plan intended to guide the growth and development of a city, town, or region, expressing official contemplations on the course its transportation, housing, and other community facilities should take, and making proposals for industrial settlement, commerce, population distribution, and other aspects of growth and development toward an ultimate state of development. There is no unified agreement on its precise meaning or use in either theory or practice. The term "master" has connotations of grandiose schemes, an ultimate state and inflexible authority. This term has been so loosely, and often incorrectly (or dishonestly), used that it has gained disfavour among planners. They now prefer to instead use the term "general plan" in the titles of their more long-range plans and proposals, because it has connotations of flexibility and thus more accurately reflects the nature of such plans. It is also used to denote a functional class of a comprehensive plan, for example, master plan for highways and thoroughfares, or those for parks and recreation.

Matrix 1. A rectangular array of rows and columns of mathematical elements that can be combined, for example, to form sums and products.

2. Something resembling a mathematical matrix, such as a list of categories along

vertical and horizontal axes with a designation of the interactions of components at the points of intersection. Matrices may be manipulated in total in a manner similar to the algebraic manipulation of single numbers, but knowledge of special rules, called matrix algebra, is necessary for such manipulation.

Matrix, Flow A matrix that indicates the existence, magnitude, and direction of flows between the boxes of a box and arrow diagram.

Mature Tree A tree that has reached the desired size or age for its intended use. Size or age will vary considerably depending on the species and intended use.

Maximax A method of choosing among alternatives by noting the attribute with the highest value for each alternative, and comparing these attributes across alternatives to choose the alternative with the maximum highest value. In order to determine the attribute with the highest value for each alternative and follow this decision rule, a very high degree of comparability between attributes within each alternative and among alternatives is needed.

Maximin A method of choosing among alternatives by noting the attribute with the lowest value or quality standard least met for each alternative, and selecting the alternative with the most acceptable lowest valued attribute or standard. In order to determine the lowest valued attribute or worst met standard, and follow this decision rule, a very high degree of comparability between attributes within each alternative and among alternatives is needed.

Maximum Modification A visual quality objective (VQO) meaning human activity may dominate the characteristic landscape. However, when viewed as background, the visual characteristics must be those of natural appearance within the surrounding area.

MBF (Thousand Board Foot Measure) A symbol to indicate 1,000 board feet of wood fiber volume either in log form or after conversion into lumber. One board foot is 1 foot x 1 foot x 1 inch or 0.144 cubic inches.

Meadow 1. An herb-land used primarily for hay production.

2. Openings in forests and grasslands of exceptional productivity in arid regions, usually resulting from high water content of the soil, as in stream side situations and areas having a perched water table.

Meadow Site Class Sites of the same dominant species and similar proportions.

Meadow SiteClass Measow sites of the same dominant species and similar proportions of species.

Meadow Sites Areas of relatively homogenous species composition that are visually significantly different from adjacent areas.

Mean Annual Increment and Culmination of Mean Annual Increment 1. The average annual growth of ***a tree,*** calculated by dividing the total growth it has accrued over its life by its age in years at the time of measurement;.

2. The total increment of increase of volume of ***a stand*** (standing crop plus thinnings) up to a given age divided by that age. Culmination of mean annual increment is the age in the growth cycle of an even-aged stand at which the average annual rate of increase of volume is at a maximum. In land management plans, mean annual increment is expressed in cubic measure and is based on the expected growth of stands.

Measurement system integrity The tracking and documentation over the long term of all causes of error or uncertainty in a final data-analysis product. These include instrument calibration, adequacy of measurement validation, data coverage and sampling density, availability and quality of ancillary data, procedures for data analysis and reduction, the results of checks against independent measurement, and quantitative error analysis.

Measurement validation The establishment of confidence in the numerical relationship between the calibrated sensor output and the actual variable being measured.

Median Lethal Dose (LD50) The amount or concentration of a toxic substance which will result in the death of 50 percent of a group of test organisms upon exposure (by ingestion, application, injection or in their surrounding environment) for a specified period of time.

Median Tolerance Limit (TLm) The concentration of some toxic substance at which just 50 percent of the test animals are able to survive for a specified period of exposure.

Memorandum of Understanding An agreement with another state or central agency making land available for occupancy or use under specified conditions.

Mensuration, Forest A science dealing with measuring volume, growth and development of individual trees and stands and determining various products obtainable from them.

Mercator projection A method of making maps in which the Earth's surface is shown as a rectangle with the meridians as parallel straight lines spaced at equal intervals and the parallels of latitude as parallel straight lines intersecting the meridians at right angles. Areas away from the equator appear larger than they are, with the greatest distortion near the poles.

Merchantable Height The point on a tree stem at which diameter limit requirements for a centain product are not met. Limits are: the point at which a sawlog tree is less than 5 inches in diameter, measure inside the bark (dib); a pulpwood tree less than 4 inches dib; or the point on any tree at which a defect is found that cannot be processed out.

Merchantable Timber A tree or stand of trees which may be disposed of at a profit through conversion to salable products.

Mesic Environmental conditions that are medium in moisture supply.

Mesopause The upper boundary of the mesosphere where the temperature of the atmosphere reaches its lowest point.

Mesosphere The atmospheric layer above the stratosphere, extending from about 50 to 85 kilometers altitude. The temperature generally decreases with altitude.

Metadata Information describing the content or utility of a data set. For example, the dates on which data were procured are metadata.

Meteor The former Soviet Union's series of polar-orbitmg weather satellites. The Meteor satellites transmit images in a system compatible with the NOAA polar-orbiting satellites.

Meteorology Study of the atmosphere and its phenomena. Study of the Earth's atmosphere, including its movements and other phenomena, especially as they relate to weather forecasting.

Metsat Generic term for meteorological (weather) satellites.

MHz (megahertz) 106 hertz.

Microclimate 1. The local climate of a site or habitat varying in size from a tiny crevice to a large land area, but being usually characterized by considerable uniformity of climate over the site involved and relatively local as compared to its enveloping macroclimate from which it differs because of local climatic factors (such as elevation and exposure).

2. A local climatic condition near the ground resulting from modification of relief, exposure, or cover.

3. The fine climatic structure of air space which extends from the very surface of the Earth to a height where the effects of the immediate character of underlying surface no longer can be distinguished from the general local climate (i.e., mesoclimate or microclimate). The microclimate varies with and in turn is superimposed upon the larger-scale conditions. While some rigid limits have been placed on the thickness of the layer concerned, it is realistic to consider variable thicknesses, e.g., the microclimate of a putting-green versus that of a redwood forest. Generally, four times the height of surface growth or structures defines the level where microclimatic overtones disappear. The word has been used to describe the climate of a cave, a plant, and an insect inside of a log.

Microeconomics Economic studies of particular individuals or single commodities; for example, the demand for wheat or for employment in the auto industry.

Microenvironment (Microhabitat) A small or restricted set of distinctive environmental conditions that constitute a small habitat, such as a tree stump, a dead animal, or a space between clumps of grass.

Micrometer (micron) One millionth of a meter; used to measure wavelengths in the electromagnetic spectrum.

Microprocessor Controlling unit of a microcomputer; laid out on a tiny silicon chip and containing the logical elements for handling data, performing calculations, carrying out stored instructions, etc.

Microrelief (Microtopography) Minor ground surface irregularities which, when compared from a fixed point of elevation between them, display differences in elevation of not more than 1 foot within a distance of 4 feet.

Microwave Electromagnetic radiation with wavelengths between about 1000 micrometers and one meter.

Middle Ground (Visual Distance Zone) Part of a scene or landscape which extends from the foreground zone 1/4 to 1/2 mile up to 3 to 5 miles from the observer. Texture, shape, and some detail are discernible at that distance. It is within this range that the emergence of shapes and patterns on the landscape can be clearly seen. Consequently this portion of a view often best shows whether human-made changes rest easily or uneasily on the landscape.

Middle infrared Electromagnetic radiation between the near infrared and the thermal infrared, about 2-5 micrometers.

Mid-range Planning (Middle Term Planning, Mid-term Planning) Planning that combines elements of both long-range planning and short range planning, typically between 10 and 25 years in the future. Often, however, the planning process is merely divided into the short-range and long-range planning with a single dividing point, and the mid-range classification is not used.

Migration Corridor Situations, usually linked to topography and vegetation, that provide a completely or partially suitable habitat that animals move through during migrations.

Milacre Plot A sample plot of 1/100O acre (usually 1/10 chain square) used in reproduction or vegetation surveys.

Millibar (mb) One thousandth of a bar; a unit of atmospheric pressure. The average atmospheric pressure at sea level is 1.01325 bars or 1013.25mb.

Millwood Timber that is used by turning mills and specialty hardwood plants.

Mine Drainage Any water forming on or discharging from a mining operation.

Mine Drainage, Acid Any acid water draining or flowing on, or having drained or flowed off, any area of land affected by mining.

Mine Tailings The waste material or refuse separated and remaining after raw minerals or ore have been processed. Mining waste from open-pit mining is termed "spoils" while "tailings" is usually used for the waste from hard rock mining.

Mineral A substance is "mineral" under the general mining laws: 1. If it is scientifically recognized as such,.

2. If it is classified commercially as such or.

3. If it derives from the Earth and possesses economic value and utility aside from the agricultural purposes of the surface itself.

Mineral Entry The filing of a mining claim for public land to obtain the right to any minerals it may contain.

Mineral Soil A soil consisting predominantly (80% or more) of and having its properties determined predominantly by inorganic matter. The soil may have an organic surface layer up to 30 centimetres thick.

Mineral, Common Variety Earth materials which, although maybe having value for use in trade, manufacture, the sciences, or in the mechanical or ornamental arts, do not possess a distinct, special economic value for such use over and above the normal uses of the general sum of such deposits. For example, sand and gravel are frequently present in abundance in most areas and that makes them a "common variety mineral". Neither mining claims nor mining claim patents can be filed for "common variety mineral" deposits.

Mineral, Excepted Types of minerals that are exempted from the regulations in the general mining laws which authorize prospecting for and development of mineral resources on public lands. The excepted minerals are: oil, gas, oil shale, coal, potassium, sodium and phosphates and in Louisiana and New Mexico, sulphour. Similarly excepted as to mineral locations made after 1955 are the common variety minerals sand, stone, gravel, pumice, pumicite, or cinders.

Mineral, Locatable Precious or semi-precious minerals that are not considered to be common variety minerals, i.e., not such materials as sand or gravel. Locatable mineral deposits can be claimed and the mining claim patented, thus converting it to private ownership.

Mineralization The breakdown of organic compounds to their inorganic (i.e., mineral) forms, e.g., proteins to nitrates, phosphates, etc.

Minimum Viable Population For planning purposes, a population having the estimated numbers and distribution of reproducing individuals, to ensure its continued existence throughout its existing range in the planning area.

Mining Claim (Claim) One unit of the public mineral lands which a miner (claimant or association), takes for mining purposes, and holds in accordance with mining laws.

Mining Claim (Patented) A mining claim to which a patent has been secured from the government by compliance with the laws relating to such claims. The patent is a legal document that conveys the title to the ground (i.e., ownership) to the claim's owner. No further annual assessment work need be done, but property taxes must henceforth be paid.

Mining Claim (Valid) A claim where there is an actual discovery of valuable minerals in quantities such that it would be profitable to mine.

Mining Debris, Hot Refers to materials in the overburden, mining refuse, mine tailings or mining spoils piles that are highly acid and difficult to re-vegetate.

Mining Debris, Sweet Refers to materials in the overburden, mining refuse, mine tailings or mining spoils that are neutral or slightly alkaline and capable of supporting certain calcium-demanding plants.

Mining Refuse The solid waste from a mineral preparation or refining plant.

Mining Spoils (Mining Waste) The overburden non-ore material removed in gaining access to the ore of mineral material in mining. Contrast with "mining refuse" which is the solid waste left by mineral processing operations. Mine "tailings" is also sometimes used in this latter sense.

Mining, Area Surface (Area Strip Mining) A type of open-pit mining or strip mining that is generally practiced on gently rolling to relatively flat terrain on relatively large tracts of land. A trench is first cut through the overburden to expose the deposit of mineral or ore to be removed. The first cut is extended to the limits of the property or deposits. The overburden from the first cut is placed on un-mined land adjacent to the cut. The mineral or ore is then removed. Once the first cut is completed, a second cut is made parallel to the first, and the overburden from the succeeding cuts is deposited in the cut just previously excavated. The final cut leaves an open trench equal in depth to the thickness of the overburden and the mineral bed removed, bounded on one side by the last spoil pile and on the other by the undisturbed highwall. The final cut may be up to a mile or more from the starting point, and the overburden from the cuts, unless graded or levelled, resembles a ploughed field or the ridges of a gigantic washboard.

Mining, Auger A mining method often used by surface mine operators when the amount of material overlying the seam from which a mineral (or ore) is being extracted gets too thick to be removed economically. Large diameter (2 to 7 feet), closely spaced holes are drilled up to 200 feet into the seam using an auger type bit whose head breaks up the deposit and whose screw-like extensions carry the material back out of the bore hole and into the open. Auger mining is usually associated with contour strip mining. It is a common practice used to recover additional tonnage after the coal to overburden ratio has become too small to render further contour strip mining economical. When the slope is too steep for contour mining, augering is often performed directly into the hillside from a narrow bench. Augers are also used to recover coal near an outcrop that cannot be extracted safely by underground mining.

Mining, Contour Surface (Contour Strip Mining) A type of strip mining that is practiced in areas of hilly topography when the coal seam is exposed on or approaches the surface at approximately the same elevation along a hill side. The conventional method of mining consists of ıemoving the over burden from the mineral seam, starting at the outcrop and proceeding around the hillside. The cut appears as a contour line, thus, the name. Overburden is cast down the hillside and stacked along the outer edge of the operating bench. After the uncovered seam is removed, successive cuts are made until the depth of the overburden becomes too great for economical recovery of the coal. Physical limitations of equipment reach, capacity, etc., may also determine the strippable limit or cut-off point for mining. Contour mining creates a shelf or bench on the side of the hill. On the inside it is bordered by the highwall, ranging in height from a few feet to more than 100 feet; and on the outer side the pit is bordered by a high ridge of spoil with a precipitous downslope that is subject to severe erosion and landslides. Certain laws require some of these slopes to be stabilized in "return to contour" restoration activities.

Mining, Lode The mining of a valuable mineral which occurs as a tabular deposit between definite, contrasting mineral or rock boundaries. "Lode", as used by miners is nearly synonymous with the term "vein" or "ore body" as employed by geologists. A lode consists of several veins spaced closely enough together so that all of them, together with the intervening rock, can be mined as a unit.

Mining, Open-pit (Opencut Mining, Open Cast Mining) A form of mining operation which extracts minerals that occur near the surface from an open excavation. Waste materials overlying the minerals are first removed and then the mineral materials are broken up and removed. The mining of metal bearing ores by surface mining methods is commonly termed "open-pit mining" as distinguished from the strip mining of coal and the "quarrying" of other non metallic materials such as limestone, building stone, etc.

Mining, Pit Surface mining, developing a quarry, or excavating whereby the material of value is removed from below the surrounding land surface.

Mining, Placer The extraction of valuable heavy minerals from a placer deposit (a mass of sand, gravel or other similar alluvial material) by concentration in running water. When water under pressure is used to break down placer deposits, placer mining is generally termed "hydraulic mining".

Mining, Strip Any operation in connection with prospecting for, excavating, or mining minerals which results in a large-scale surface or stream bottom disturbance from stripping, trenching, dredging, rim cutting or open-pit digging. There are two general types: area and contour mining.

Mining, Surface (Strip Mining, Open Cast Mining, Open-cut Mining, Open-pit Mining) A very broad term referring to any process of removing the earth, rock, and other strata in order to uncover the underlying mineral or fuel deposit.

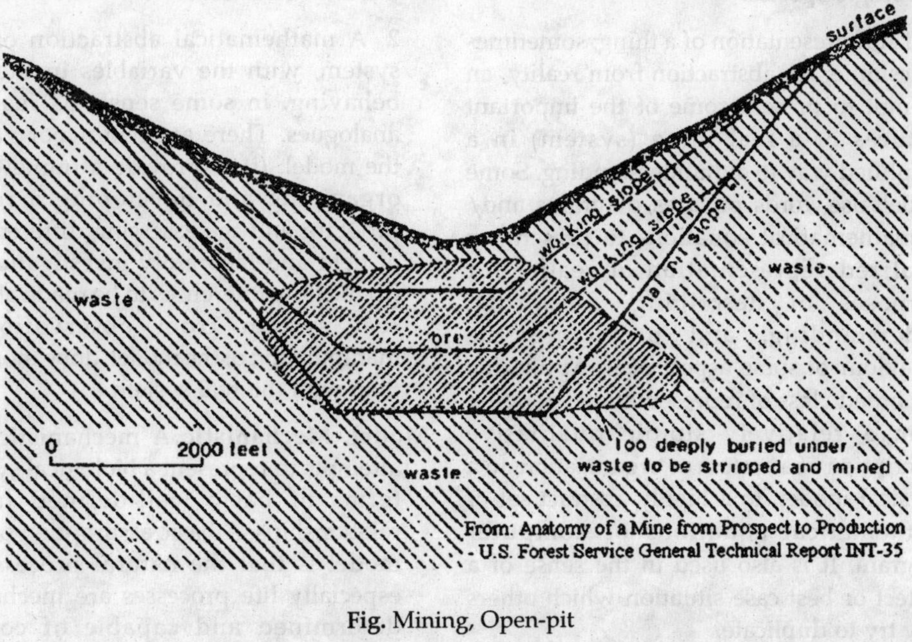

Fig. Mining, Open-pit

Miscellaneous Land Type A mapping unit for areas of land that have little or no natural soil, or that are too nearly inaccessible for orderly examination, or that occur where, for other reasons, it is not feasible to classify the soil. Examples are alluvial land, bad lands, made land, marsh, mine dump, mine wash, river wash, rock land, rough broken land, rubble land, scoria land, swamp, secure military areas, and urban land.

Mitigate To reduce the impact of an action by moderating its force, intensity, or scope.

Mitigation Actions to avoid, minimize, reduce, eliminate, or rectify the impact of a development or management practice.

Mitigation, Impact Measures which are taken to minimize the significant potential environmental impacts of a given development alternative on various environmental factors, or the measurable reduction or elimination of an effect.

Mixed Stand A stand in which less than 80 percent of the trees in the main canopy are of a single species.

Model A representation of a thing; sometimes a facsimile. An abstraction from reality, an attempt to present some of the important features of a real thing (system) in a simplified way to aid understanding. Some models use words, pictures, diagrams, and/or mathematical equations to present an idealized representation of reality for purposes of describing, analyzing, understanding, and predicting the behaviour of some aspect of it. Applicable to a broad class of representations, ranging from a relatively simple qualitative description of a system or organization, to a physical model, to a highly abstract set of mathematical equations or computer program. It is also used in the sense of a perfect or best case situation which others may try to duplicate.

Model, Conceptual A mind's-eye perception of what something is; the image conjured up by prompting from both the physical and theoretical aspects of the entity of concern, whether that entity be a simple object, an abstract system, or a multidimensional problem.

Model, Deterministic A model in which the state of the system results from determining causes, and those causes can be identified and adequately described without considering probabilistic elements. Each aspect or factor of the model is specified. Risk is not included.

Model, Graphical The graphic portrayal of an idea, thought, or concept e.g., a trophic level diagram of a functional ecosystem.

Model, Mathematical 1. A conceptual tool which relates the general characterization of a process, object, or concept in terms of mathematics, thus enabling the relatively simple manipulation of variables to be accomplished in order to determine how the process, object, or concept would behave in different situations.

2. A mathematical abstraction of a real system, with the variables in the model behaving, in some sense, as real-world analogues. There are two major values of the model: (i) to provide a framework for organizing and developing a research program and the assessing different facets of field and laboratory investigations; and (ii) to allow inferences to be made about the real system from observation and manipulation of the model, done more easily than with the real system.

Model, Mechanistic A mechanistic model attempts to represent a biological system in terms of basic well-defined laws or relationships. A basic assumption of such a model is that the natural processes and especially life processes are mechanically determined and capable of complete explanation by the laws of physics and chemistry.

Model, Probabilistic A model in which most of the key variables may take on more than one value which may be assigned based on probable distributions. It allows randomness to be used or studied. Such models are sometimes called stochastic which means, litreally, making a best guess.

Model, Simulation A dynamic mathematical model that mimics the functioning of a system or process by step-by-step solutions of the equations which describe the system. Such models often allow the consequences of different inputs and processes to be evaluated as to the effects that they may have on the system outputs.

Model, Stochastic (Probabilistic Model) 1. A mathematical model the variables of which (all or some) are drawn at random from some specified distribution.

2. A model that attempts to include the effects of random variability in driving variables and parameters.

3. A stochastic model is one that does not offer a unique solution but rather a distribution of solutions with the probability of each solution corresponding to some specified probability distribution.

Model, Systems Conceptual representations of an ecosystem of concern... not all-inclusive but rather, all encompassing. That is, they include abiotic, producer, consumer, decomposer, and nutrient components, but they may not include all of the interesting or important biological processes or parts that are found in the ecosystem. The specific miniaturization, omissions, or substitutions depend upon the objectives for the modeling activity.

Modification (VQO) A visual quality objective meaning that people's activity may dominate the characteristic landscape but must, at the same time, utilize naturally established form, line, colour, and texture, so that its visual characteristics are those of natural occurrence with the surrounding landscape.

Modified Cut Areas Areas that will require modification of normal timber regeneration methods (leading to even-aged stands) because of landscape management quality standards or filter strip criteria. Depending on conditions in each area, cutting methods will be used which may not result in even-aged management. Variance may range from no cutting to shelterwood, selective cut, or group selection cuts.

Modified Cutting An even-aged harvest cutting method modified to the extent that it no longer meets the definition of that even-aged harvest cutting method, but will insure the perpetuation of the desired commercial tree species. Use is limited to special management situations where normal mitigating and/or coordinating measures will not provide the desired effects for resource concerns other than timber. The selection of trees or groups of trees to be harvested is determined by a silviculturist through analyzing stand and site conditions in concert with the site specific objective for which the modified cutting is done.

Modulation Modification of any aspect of a land use plan in order that it more nearly expresses the constituency's interpretation of the relevant goals and objectives. An adjustment of parts in proportion or relative to the whole.

Monitoring and Evaluation Evaluating a sample of management practices to determine how well objectives have been met, as well as the effects of those management practices on the land and environment.

Monitoring Program, Environmental A program for measuring anticipated disturbances in environmental systems to assure that they do not exceed acceptable limits. The program often includes certain aspects of a baseline study program selected for their ability to detect alterations in the ecosystem which are induced by a project.

Monitoring Systematically collecting information to evaluate changes in actions, conditions, and relationships over time and space or progress toward meeting desired conditions or plan objectives.

Monsoon Heavy winds characterized by a pronounced seasonal change in direction. Winds usually blow from land to sea in the winter; while in the summer; the flow reverses and precipitation is more common. Monsoons are most typical in India and southern Asia.

Montreal Protocol An international agreement to drastically reduce CFC production, the Protocol was adopted in Montreal in 1987. It was significantly strengthened at a subsequent meeting in London in 1990 that called for a complete elimination of CFCs by the year 2000. The agreement was again amended by a Meeting of the Parties in Copenhagen in November 1992. Consumption of controlled substances—such as CFCs and halons—was greatly reduced or eliminated, and many accountability dates were moved forward, often from January 2000 to 1 January 1996.

Morphological Map A morphological map represents the form of the surface of the land, e.g., a map of occurrence of various slope steepness categories.

Morphology, Land Land "morphology" is restricted to studies of the surface form of the Earth without any reference to the processes responsible for those forms.

Mosaic A composite picture built up from a number of image segments. An example of a mosaic is the WEFAX transmission, which includes both polar and mercator mosaics derived from TIROS-N/NOAA polar orbit image data.

Most Probable Number (MPN) A statistical evaluation of degree of water pollution based on presence of coliform bacteria. It is not feasible to identify the exact concentration of coliform bacteria in a water sample. The MPN interprets test results in terms of results observed.

Mountain and valley breezes System of winds that blow downhill during the night (mountain breeze) and uphill during the day (valley breeze).

Mountain Top Removal In this mining method, 100% of the overburden covering a coal seam is removed to recover 100% of the mineral. Excess spoil material is hauled to a nearby hollow to create a valley fill.

Muck Highly decomposed wet organic matter in which the original plant parts are not recognized.

Muck Soil A wet soil containing between 20% and 50% organic matter.

Mulching Providing any loose covering for exposed forest soil, using organic residues, such as grass, straw or wood fibres to protect exposed soil and help control erosion.

Multidisciplinary Team Two or more people having different training and/or backgrounds all assembled and assembled and assigned responsibilities in the same activity or effort. Each specialist is assigned a portion of the problem and his or her partial individual solutions are linked or combined to provide the complete solution. No one discipline is sufficiently broad to solve adequately the problem. This is different from an "interdisciplinary team" which does not break the problem apart by disciplines, but instead works with frequent interaction so that each discipline may provide insights into any part of the problem and so that disciplines may combine to provide new solutions beyond the scope of any single discipline.

Multiple Use A term and concept of long-standing debate. At least three different ideas are involved: 1. Different uses of adjacent sub-areas which together form a composite multiple-use area,.

2. The alternation in time of different uses on the same area, and.

3. More than one use of an area at one time. In the first two ideas it is implicit that direct competition between uses is avoided by alternating them in space or in time. The third idea involves simultaneous use of one space and must deal with satisfactory resolution of conflicting activities, and incompatible uses by different groups of different size and economic, social, and political influence.

Where spatially coincident uses are involved at a given time, conflicts between resource users will almost always occur and the concept of such forms of multiple use should be realistically interpreted as a dominant use with secondary uses integrated only in so far as they are compatible with the first. However, where the idea of incompatibility relates to whether the productivity or yield of a single resource has been maximized, multiple use can perhaps be validated in terms other than single-resource production efficiency. To the single purpose resource user, multiple-use may seem inefficient or economically ruinous. Under this concept the aim of resource use allocation is to maximize the national well being, promoting general social and economic prosperity. Social needs are not necessarily best served by maximizing the production of a single resource (or even by maximizing the production of several resources) but by an over-all mix of total national resource uses that brings the greatest social and economic benefits.

Multiplexer A device that combines several separate communications signals into one and outputs them on a single line.

Multispectral Scanner (MSS) A line-scanning instrument flown on LANDSAT satellites that continually scans the Earth in a 185 km (100 nautical miles) swath. On LANDSATs 1, 2, 4, and 5, the MSS had four spectral bands in the visible and near-infrared with an IFOV of 80 meters. LANDSAT-3 had a fifth band in the thermal infrared with an IFOV of 240 meters.

N

Nadir Point on Earth directly beneath a satellite, the opposite of zenith. Compare with subsatellite point.

Nanometer (nm) One billionth of a meter. Nanometers are used to measure wavelengths in the electromagnetic spectrum.

Native Bacteria Bacteria indigenous to natural water bodies.

Native Organism Animals or plants that originated in the area in which they are found, i.e., were not introduced and naturally occur in that area.

Native Species 1.Those plants and animal species indigenous to the plan area or assessment area.

2. Species indigenous to an area of analysis.

Natural Appearing A landscape that appears to have evolved to its present condition as a result of the forces of nature. Human-made changes are present, but appear to be the result of natural change.

Natural Area Preserve An area which retains a natural or relatively natural conditions and exemplifies typical or unique vegetation and associated biotic, soil, geologic and aquatic features. The unit is maintained in natural conditions by allowing physical and biological processes to operate, usually without direct human intervention. Within it, wild parks, sanctuaries, refuges or wilderness can be designated.

Natural Area, Research (Natural Area Preserve) An area designated by a public or private agency specifically to preserve a representative sample of an ecological community, primarily for scientific and educational purposes. Commercial exploitation is ordinarily not allowed and general public use is discouraged. Where the vegetation in the Natural Area is not of a stable, climax type, management measures may be taken to keep it at its present state, i.e., prescribed burning or controlled grazing to conserve certain floral and/or faunal characteristics. Research and study that is nondestructive and consistent with the purpose for which such an area has been established is encouraged. The general public may be excluded or restricted where necessary to protect studies or preserve these areas. Designated areas representing as many as possible of the major, natural timber types or other plant communities in unmodified condition. Other forest or range conditions that have special or unique characteristics of scientific or educational interest, such as examples of grass or timber types near the limits of their environmental range, unique bog associations, or unusual combinations of flora may also be set aside. To qualify as a Natural Area Preserve, a site should be under protection, usually by a nonprofit agency or a Government agency. To whatever extent is feasible, animal life should also be present in unmodified condition. Research Natural Areas should be large enough to provide essentially unmodified conditions in their interior portions — usually over 300 acres. Exceptions to the usual minimum of 300 acres should be limited to truly outstanding cases. Research on Natural Areas is nondestructive in nature.

Natural Forest The condition of a forest environment at any point in time, including its associated plant and animal communities, which has been reached essentially through the process of natural disturbance regimes (catastrophic occurrences) and natural ecological succession.

Natural Hazard (Environmental Hazard) Those elements in the physical environment, harmful to people and caused by forces extraneous to them.

Natural Hazard Lands Areas where uncontrolled or incompatible development could unreasonably endanger life and property. These lands include flood plains and areas frequently subject to weather disasters, areas of unstable geological, ice, or snow formations, and areas with high seismic or volcanic activity.

Natural History Resource Natural phenomena which reference the development of the Earth's surface and the evolution of life. Two interrelated categories of natural features are recognized: 1. The geological category, with results from forces and processes acting on the Earth's surface to produce land forms and other nonliving entities; and.

2. The ecological category, with living entities and processes between biological forms and their environments.

Natural Regeneration Naturally establishing a new crop of trees on an area following timber harvest by seed fall from trees left for that purpose, by hardwoods sprouting from roots or stumps, or by seed or seedlings already in place.

Natural Resource A feature of the natural environment about what choices must be made that is of use and value in serving human needs. It need not be possessable to be an object of choice. These are things such as air, water, sunshine, wild animals, or wild vegetation. The term "natural resources" may also encompass the substances, forces, conditions, relationships, and other aspects of nature which are transformed by people and underlie, shape, or affect that complex mixture of natural and cultural landscapes which constitute the environment.

Natural Selection A process (with respect to a particular trait) in which, if , a population has (a) variation, (b) fitness, and (c) a heritable component, then the distribution of the trait will differ among life history stages, and if the population is not in equilibrium, the trait distribution will differ between parents and off spring more than would be expected from conditions a and c. (*from John Endler* ***Natural Selection in the Wild***).

Natural Vegetation The vegetation condition and/or species composition native to an area prior to it being modified by people. Little existing vegetation is unmodified by people's activities either directly (i.e., burning or cutting) or indirectly (i.e., introducing grazing animals or exotic plants). The current tendency is to call all vegetation not deliberately managed or controlled by farming activities "natural vegetation" or to use the term "semi-natural." The phrase rarely includes the mosses, lichen, algae, etc.

Nautical mile A unit of distance equal to exactly 1.852 kilometers or about 6076.1 feet. A nautical mile is approximately equal to 1/60 of a degree or 1 minute of arc of a great circle of the Earth (i.e., 1 minute of arc of latitude or of longitude at the equator).

Near infrared Electromagnetic radiation with wavelengths from just longer than the visible (about 0.7 micrometers) to about two micrometers.

Need for Change A finding by the Responsible Official that there is a need to modify plan components through a review of new issues and information, monitoring and evaluation results, and changes in law or regulation.

Needs The fundamental motivations and requirements of people, not to be confused with "wants" or "desires". The distinction between "needs" and "wants" becomes subject to debate beyond the minimal levels of basics such as food, clothing, shelter, drinkable water, and breathable air. Items that most people would consider necessities of life may be useless extravagances to other cultures. Psychological studies have indicated that conditions that should have no effect on the body may be necessities for maintaining mental health, and therefore physical health and productivity.

Negative Declaration Statement (Negative Declaration) A finding that a proposed action on the land does not constitute a major central, state, or local agency action significantly affecting the quality of the human environment and therefore does not require an environmental impact statement.

Nekton The whole group of macroscopic aquatic animals that swim actively and may move long distances for feeding or breeding. Aquatic organisms that are carried passively by water currents or the wind are termed plankton and those that dwell on the bottom benthos.

Neoteric Area Sites and areas which contain outstanding examples of people's modern culture which will likely become historic properties in the future.

Nephanalysis A type of analysis using satellite cloud pictures to study the relationship between cloud forms and storm systems. In classical mythology Nephele was a woman Zeus formed from a cloud.

Nepheloccygia Clouds that resemble recognizable shapes.

Nesting The behaviours of adult animals leading up to nest building and egg laying through departure of young from the nest.

Net Public Benefit The overall long-term value to the nation of all goods and services, outputs, and positive effects (benefits) less all associated inputs and negative effects (costs) whether they can be quantitatively valued or not. Net public benefits are measured by both quantitative and qualitative criteria rather than a single measure and index.

Net Value Change (Also Net Resource Value Change) The sum of the changes resulting from increases (benefits) and decreases (damages) in the value of outputs from the land area, e.g., the consequences of fire or flood.

Net Worth Net monetary value.

Newton's law of universal gravitation All bodies attract each other with what is called gravitational attraction. This applies to the largest stars as well as the smallest particles of matter The force of attraction between two small bodies (or between two spherical bodies of any size) is proportional to the product of their masses and inversely proportional to the square of the distance between their centers. In other words, the closer two bodies are to each other; the greater their mutual attraction. As a result, to stay in orbit, a satellite needs more speed in a low than a high orbit.

Kepler's three laws of planetary motion, which had been derived empirically by Johannes Kepler; were obtained with mathematical rigor as a consequence of Newton's law of universal gravitation in conjunction with his three laws of motion.

Niche A place or position of life suitable for the capabilities or merits of an organism or qualities of a thing. A site or habitat supplying the factors characteristically necessary for the successful existence of an organism or species. An "ecological niche" is the role of an organism in an ecological community, involving especially its way of life and its effect on the environment, as through its relations to other biotic and abiotic factors. It also may mean where it lives, what it does (how it transforms energy, behaves, responds to and modifies its physical and biotic environment), and how it is constrained by other species. By analoguey, it may be said that the habitat is the organism's "address," and the niche is its "profession." The "ecological niche" can be subdivided into a "trophic niche" and a "multidimensional niche or hypervolume."

1. "Trophic niche" is the feeding role or position of an organism in a biotic community - e.g., producer, consumer, herbivore, carnivore, detritivore, etc.

2. The "Multidimensional niche or hypervolume" is an organism's position within a volume described by environmental gradients, such as temperature, moisture, pH, and soil factors.

3. The Forest's, Grassland's, or Prairie's role in contributing to social, economic,and ecological sustainability.

Nitrogen Cycle The sequence of biochemical changes undergone by nitrogen, wherein it is used by a living organism, liberated upon the death and decomposition of the organism, and converted to its original state of oxidation.

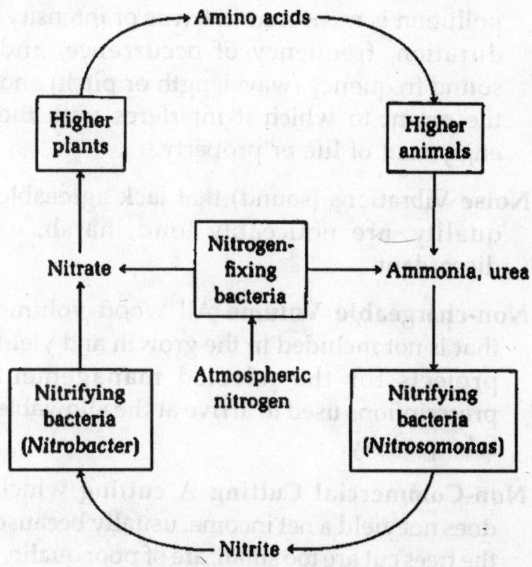

Fig. Nitrogen Cycle

No Action Alternative An alternative in a plan representing the condition expected to exist in the future if current management direction continues unchanged. Also referred to as the "current program.".

No Surface Occupancy Stipulation When included in an oil and gas lease, a condition that prohibits using the applicable land surface for access roads, drilling sites, etc.

No Surface Occupancy The prohibition against locating well sites, evaporation ponds, pits, tank batteries, and other oil and gas production facilities on a certain tract of land. Access roads and gas transmission pipelines may be located on a "no surface occupancy" area if other resource management activities would not be prohibited from similar types of development and the impacts can be mitigated.

No Till; Zero Till Planting a crop without prior seedbed preparation into sod, crop residue, or an existing cover crop and eliminating subsequent tillage operations.

Noise Pollution The addition of energy in the form of sound to the environment beyond what would naturally occur. The degree of pollution is measured in terms of intensity, duration, frequency of occurrence, and sound frequency (wavelength or pitch) and the extent to which it interferes with the enjoyment of life or property.

Noise Vibrations (sound) that lack agreeable quality, are noticeably loud, harsh, or discordant.

Non-chargeable Volume All wood volume that is not included in the growth and yield projects for the selected management prescriptions used to arrive at the allowable sale quantity.

Non-Commercial Cutting A cutting which does not yield a net income, usually because the trees cut are too small, are of poor quality, or are not marketable.

Non-Consumptive Use Using a resource in a way that does not reduce the supply, such as many types of recreation.

Non-Consumptive Wildlife All wild creatures (rarely plants) not taken by people for sport, food, or profit.

Non-declining Yield A level of timber production planned so that the planned sale and harvest for any future decade is equal to or greater than the planned sale and harvest for the preceding period, usually a decade.

Non-Forest Acres Areas in roads, rights-of-way for utilities, special-use areas, wildlife openings, lakes, etc.

Non-Forest Land Lands never having or incapable of having 10 percent or more of the area occupied by forest trees, or lands previously having such cover and currently developed for non-forest use. Land where timber utilization is precluded for other uses. Includes areas used for crops, improved pasture, residential areas, improved roads of any width and adjoining clearings, and power line clearings of any width. If intermingled in forest areas, unimproved roads and non-forest strips must be more than 120 feet wide, and clearing more than one acre in size to qualify as non-forest land.

Non-Game Species Animal species that are not usually hunted or may not be hunted in a state. This classification is determined by the State Legislature.

Non-Industrial Private Forest Land (NIPE) Forest land owned by a private individual, group, or corporation not involved in wood processing.

Non-Market Values Estimated values for goods and services valued in terms of what reasonable people should be willing to pay rather than go without the good or service but not traded in dollars and cents, e.g., fish and wildlife values, scenic quality values.

Non-Persistent Pesticide A pesticide the harmful effects of which are of relatively short duration and, therefore, do not normally contaminate the environment for long periods after application. Phosphate-based Malathion is an example of a non-persistent pesticide.

Non-Point Source Pollution Pollution arising from all ill-defined and diffuse source, such as runoff from cultivated fields, grazing land, or urban areas.

Non-Point Source Pollution Pollution, the source of which is general rather than specific in origin (e.g., sediment from road construction, logging operations, or bacteria from swimming areas).

Non-Renewable Resources Resources the total quantity of which physical quantity does not increase significantly with time. Thus, with the total initial supply being limited in quantity, each use must diminish the total stock. Long before a given resource is physically used up, it may be "exhausted" in the sense that further utilization is indefinitely discontinued because the costs of producing any possible quantity of this resource are larger than the revenue that could be obtained. The economic concept of resource exhaustibility typically only applies to commodity type resources because biological, location, and amenity type resources can be physically (by extinction or site occupancy) or conceptually (alteration of primeval wilderness conditions) exhausted.

Non-selected Roadless Area A roadless or undeveloped portion of a National Forest greater than 5,000 acres (or smaller than 5,000 acres but contiguous to designated wilderness or primitive areas) which, after administrative review, has been deleted from consideration for inclusion in a wilderness area system.

Non-structural Range Improvement A modification of existing vegetation to improve the grazing resource. Examples are spraying or plowing sagebrush and seeding the treated area to grass.

Non-Substantive Changes A modification to the plan that does not appreciably change plan components (e.g.,desired conditions, objectives, guidelines).

Norm An ideal standard of conduct or ethical value binding upon the members of a group which serves to guide, control, or regulate proper and acceptable behaviour.

Normative Of, relating to, or dealing with norms, i.e., the standards of conduct or ethical value that are binding upon the members of a group and serve to guide, control, or regulate proper and acceptable behaviour. Normative variations exist even in a relatively stable culture. These variations are due to the generalized nature of norms relative to specific social situations and to the variations in their perception and interpretation by individuals. Moreover, in societies, where structure manifests sub-cultural variations, ambiguities and variations about the meaning of norms are to be expected. Norms are external to the individual to the extent that normative definitions exist in the reality defined by a particular culture, which may pre-exist his or her being. There is a hierarchy of such definitions of "right and wrong". These are traditionally identified as "folkways", mores, and laws.

Noxious Species An undesirable plant species that is unwholesome to range or animals. (Not to be confused with species declared noxious by certain laws.).

Noxious Weed A noxious, destructive, or troublesome plant when found to be in epidemic proportions and of negative economic importance.

Nursery Areas Areas (early summer range) used by a temporary big game social unit consisting of cows and young calves.

Nutritive Value Relative capacity of a forage to furnish nutrition for animals.

O

Oasis A spot in a desert made fertile by water; which normally originates as groundwater.

Objective Function The equation (or set of equations) expressing the quantity to be maximized, stabilized, or minimized in a system along with the constraints.

Obligations Amounts of orders placed, contracts awarded, services rendered, or other commitments made by landowners or agencies during a given period which will require outlays during the same or some future period.

Oblitreation Returning the land occupied by a road or trail to production.

Observer Normal That visual relationship between viewer and viewed which exists when the location from which something is seen is at (or approximately at) the level of that object or the dominant visual elements in the surrounding landscape.

Observer Superior That visual relationship between viewer and viewed which exists when the location from which something is seen is above the level of that object or the dominant visual elements in the surrounding landscape. From the "observer superior" position, the maximum extent and expanse of view is expected to be visible with atmospheric conditions and earth curvature being ultimate distance limitations.

Occluded front (occlusion) A composite of two fronts formed as a cold front overtakes a warm front. A cold occlusion results when the coldest air is behind the cold front. The cold front undercuts the warm front and, at the Earth's surface, coldest air replaces less-cold air.

A warm occlusion occurs when the coldest air lies ahead of the warm front. Because the cold front can not lift the colder air mass, it rides piggyback up on the warm front over the coldest air.

Occupancy Trespass The illegal occupancy or possession of a tract of land.

Ocean The salt water surrounding the great land masses. The land masses divide the ocean into several distinct portions, each of which also is called an ocean. The great body of salt water which occupies two-thirds of the surface of Earth, or one of its major subdivisions. The oceans include the Pacific Ocean, the Atlantic Ocean, the Indian Ocean, and the Arctic Ocean.

Odour Odour in water can be caused by a variety of materials, both natural and foreign, in the water body. Odour tests are made by using the human sense of smell, whereby a panel of "testers," each person in turn, sniffs dilutions of a water sample until that dilution with the least but definitely perceptible, odour to all on the panel is determined.

Off-Road Vehicle (ORV) Any motorized vehicle not greater than 40 inches in width, designed for or capable of cross-country travel on or immediately over land, water, sand, snow, ice, marsh, swampland, or other natural terrain; (e.g., motorbikes, 3- and 4-wheelers, four-wheel drives, air-cushion vehicle, and snowmobiles). Not included are: (a) any registered motor boat, (b) any fire, military, emergency, or law enforcement vehicle when used for emergency purposes, and any combat or combat support vehicle when used for national defence purposes, and (c) any vehicle whose use is expressly authorized by the respective agency head under a permit, lease, license, or contract.

Old Growth Terrestrial ecosystems characterized by vegetation and associated animals requiring the oldest and most mature successional stages. Old growth forests contain trees normally beyond the age of optimum maturity for economic timber harvest; often showing decadence; trees with their associates that are older than the rotation age, e.g., 120 years in forests; the last stage in forest succession. There are 10 criteria suggested within Guidance for deciding when land with trees may be called "old-growth.".

Old Growth Component That condition of a timber stand where the most dominant trees are 100 years old or older. A condition in which dead and decaying material is present in both standing trees and debris on the forest floor. It will occur in even-aged stands 100+ years old and in uneven-aged stands when most of the largest trees have reached that age, provided workers have not removed most dead and decaying material.

Old Growth Forest A forest dominated by long lived species that has escaped catastrophic disturbance for at least 120 years: it usually has large old dying trees, large snags and down logs.

Oligotrophic A lake condition characterized by a low supply of nutrients and consequent support of very little organic production. Dissolved oxygen is present at or near saturation throughout such lakes during all seasons of the year. Characteristic of waters that are poor in dissolved inorganic or mineral nutrient materials.

Omnivore An organism that feeds on both animal and plant substances. (Whether the foods must be living is debated, e.g. for "scavengers" or "detritivores.").

Open Range All suitable range within an area upon which grazing is permitted. Also, ranges which have not been fenced into management units.

Open Road Equivalents A measure of access that addresses all types of roads and trails used by motorized vehicles. Miles of secondary and primitive road are converted to equivalent primary road miles.

Open Space Land and water areas on which few or no buildings and development has been made, relatively permanent, serving the following purposes: 1. Conserving and protecting land and its resource features,.

2. Providing ecological protection,.

3. Providing park and recreation uses,.

4. Providing historic and/or scenic purposes,.

5. Shaping and guiding urban form,.

6. Enhancing community values and safety, and.

7. Maintaining options for the future.

Open to Public Travel Except during scheduled periods, extreme weather conditions, or emergencies, open to the general public for use with a standard passenger automobile, without restrictive gates or prohibitive signs or regulations, other than for general traffic control or restrictions based on size, weights, or class of registration.

Open Vegetation In habitat evaluation, all clearcuts, meadows, and other openings.

Opinion A time-bound judgment or speculation that may fluctuate. In contrast to beliefs (i.e., tenaciously held convictions) "opinions" express what people believe at a given moment. "Opinions" generally are short-run but informed impressions or "guesses" about specific objects, ideas, issues, or events that are usually involved with aspects of public affairs.

Opportunity A proposal that is considered in developing alternative activities, projects, or programs wherein an option exists to invest *profitably* to improve or maintain a present condition.

Opportunity Cost The minimum value of the benefits foregone or given up due to the effect of choosing another management alternative that either impacts existing outputs or shifts resources away from other activities so that they are no longer produced and their benefits are lost.

The value of benefits that could have been produced by the best alternative use of resources forgone or given up due to the effect of choosing another management alternative.

Optimization Usually selecting the most cost-effective combination of management practices. The process of deciding on the inputs that achieve (solve) the objective function. It is the process of achieving a maximum, minimum, or zero condition subject to one or more constraints.

Optimum Land Use That feasible use which in the opinion of the decision-maker provides the minimum difference between actual and desired conditions of the landowner or public for at least 50 years. Such use satisfies an objective (maximizes, stabilizes, or minimizes) subject to a set of constraints.

Option Value The amount an individual would be willing to pay to preserve (or would have to be paid to get him or her to sell) an option to participate in some activity or to use some resource at some future time, whether or not he or she ever actually participates or uses the resource.

Orbit The path described by a heavenly body in its periodic revolution. Earth satellite orbits with inclinations near 0° are called equatorial orbits because the satellite stays nearly over the equator. Orbits with inclinations near 90° are called polar orbits because the satellite crosses over (or nearly over) the north and south poles.

Orbital plane An imaginary gigantic flat plate containing an Earth satellite's orbit. The orbital plane passes through the center of the Earth.

Ordinal Value Numerical values assigned to a variable which represent a ranking or sequence. The numerical values have meaning only when compared to one another. Because of this, mathematical operations performed on the values (means, modes, differences, etc.) will not necessarily be valid, and the results of such operations must be carefully interpreted.

Ordinance A local law, regulation, or decree enacted by a city council or other similar body under powers delegated to it by the state.

Ordinary Planning (Operational Planning) Any type of planning which, accepting existing social institutions as given, seeks to document or communicate in some way optimal alternative strategies for attaining a stated desired future.

Organic Groups Organic chemical molecule groups such as phenols. Analytic techniques include mass spectrometry, NMR (Nuclear Magnetic Resonance) and IR (Infrared Spectroscopy).

Organic Soil A soil that contains a high percentage (greater than 25 percent) of organic matter in the upper layers (where living roots are primarily found).

Ornithology The study of birds.

Orphan Banks Earthen and rock piles near abandoned surface mines operated prior to the enactment of comprehensive reclamation laws. These typically require reclamation.

Orphan Lands (Orphan Banks) Disturbed land surfaces resulting from surface mining that were inadequately reclaimed by the operator and for which that person is no longer legally responsible.

Outcome The long-term results of a program activity compared to its intended purpose. Outcome is a state of being similar to long-term ecological, social, or economic condition or goal (such as the maintenance of an ecosystem's biodiversity, jobs and income, or the quality of a regions' surface water as measured by indicators).

Outdoor Recreation and Wilderness System A system that provides outdoor recreation opportunities and wilderness experiences. The system may include all activities necessary to 1. Develop, protect, and administer the forest recreational, scenic, and wilderness resources;.

2. Develop new knowledge to improve the effectiveness of providing and managing outdoor recreation opportunities; and.

3. Provide assistance to other landowners in developing forest-based recreational opportunities.

Output A broad term for describing any result, product, service or concern that a system produces by its activities. They are measurable and capable of being used to determine the effectiveness of programs and activities in meeting objectives. The unit of measure should indicate or serve as a proxy for what the recipients get rather than what the agency does in the process of producing the given output. Example: timber sold, recreation use, livestock grazing use, etc. Any good, service, or on-site use that is produced from rural resources.

In computer and related system use, the term means reports, graphs, or data written out or shown on a screen. The media may be paper, tape, disk, computer monitor, or may be an electronic signal (e.g., to stop a system when a critical condition is reached.) It may mean the actual or potential benefits from a system.

Market output, is a good, service, or on-site use that can be purchased at a price. ***Non-market output*** is a good, service, or on-site use not normally exchanged in a market.

Output Coefficient Values which relate an unit of land (e.g., an acre) to a particular quantity of output in a specific period of time.

Output, Induced The indirect result of system activities. Generally, the primary outputs of another resource system. An example is a specific form of wildlife habitat improvement (primary) that produces cut trees for sale (induced).

Output, Primary The main goods, services, or environmental conditions of a resource system, e.g., the production of animal-unit-months of forage for livestock or board feet of timber harvested.

Outstanding Mineral Rights Specified mineral rights that are privately owned (outstanding third party) as a result of having been separated from the surface by the prior title history of the land.

Overbrowsing Excessive use of browse usually found where there is an over-population of game. Similar to overgrazing. except that overgrazing refers to grasses and forbs, while overbrowsing refers to shrubs and trees.

Overburden Material of any nature, consolidated or unconsolidated, that overlies a deposit of useful materials, ores, or coal.

Overgrazing, Local Grazing in excess of a standard (economic, agronomic, etc.) in localized areas on a range, as, for example, near watering places.

Overland Flow (Sheet Flow) The rain storm or snow melt runoff water which flows over the ground surface as a thin layer (as opposed to channelized runoff which occurs in rills and gullies). "Overland flow" typically causes sheet erosion.

Overmature Forest A forest in which, as the result of age, growth has almost entirely ceased, and decay and deterioration has accelerated.

Overstocked The situation in which trees are so closely spaced that they are competing for resources, resulting in less than full-growth potential for individual trees.

Overstocking Placing a number of animals on a given area that will result in overuse by the end of the planned grazing period. Not to be confused with overgrazing because an area may be overstocked for a short period, but the animals may be removed before the area is over-utilized. However, continued overstocking will lead to overgrazing.

Overstory The roughly horizontal uppermost layer of mature trees which overtop other layers of foliage. The canopy in a stand of trees. In contrast to the understory which is low growing woody or herbaceous vegetation fonning a layer beneath the overstory.

Overtopped Crown Class Trees with crownsenn'rely below the general level of the crown cover receiving no direct light either from above or from the sides. Syn. Suppressed.

Overturn, Fall A physical phenomenon that may take place in a body of water during early autumn. The sequence of events leading to fall overturn include: 1. Cooling of surface waters;.

2. Density change in surface waters producing convection currents from top to bottom;.

3. Circulation of the total water volume by wind action; and.

4. Vertical temperature equality. The overturn results in a uniformity of the physical and chemical properties of the entire water mass. Fish deaths often accompany the low oxygen in the changed water layers.

Overturn, Spring A physical phenomenon that may take place in a body of water during the early spring. The sequence of events leading to spring overturn include: 1. melting of ice cover;.

2. Warming of surface waters;.

3. Density changes in surface waters producing convection currents from top to bottom;.

4. Circulation of the total water volume by wind action; and.

5. Vertical temperature equality. The overturn results in a uniformity of the physical and chemical properties of the entire water mass.

Ownership The exclusive right to occupy, sell, bequeath, or mortgage property.

Ownership, Fee Simple (Fee Simple Absolute, Fee Simple Title) Ownership that may last forever and may be inherited by all heirs of an individual owner. A term usually used in the sense of "fee simple absolute" which refers to ownership that has no limitation, qualification, or condition affecting it and is the maximum possible ownership in real estate under the system of property rights founded on English common law.

Oxygen Demand, Biochemical (Biological Oxygen Demand; BOD) 1. A measure of the demand on a water body's dissolved oxygen supply which will be generated (over some specified period of time) by the biological decomposition of various additions of organic wastes. A high BOD may temporarily, or permanently, so deplete oxygen in water as to kill aquatic life. Knowing BOD is perhaps most useful in evaluating impact of wastewater on the receiving water bodies.

2. A test for detecting and measuring pollution in which the quantity of oxygen that has been used by oxidizable materials under standardized conditions is determined.

Oxygen Demand, Chemical A measure of the amount of a water body's dissolved oxygen supply that would be used up in completely oxidizing added inorganic oxidizable compounds — such as in the oxidation of ammonia to nitrate. Biological oxygen demand-(BOD) tests can only measure the biodegradable fraction of the total potential dissolved oxygen consumption by added wastes; however, COD tests may be used to measure the oxygen demand created by toxic organic or inorganic compounds as well as by biodegradable substances. A standard COD test, therefore, can be used to evaluate many industrial type wastes not readily analyzed for water quality factors by the sewage-oriented BOD test.

Ozone An almost colourless, gaseous form of oxygen with an odour similar to weak chlorine. A relatively unstable compound of three atmos of oxygen, ozone constitutes—on the average— less than one part per million (ppm) of the gases in the atmosphere (peak ozone concentration in the stratosphere can get as high as 10 ppm). yet ozone in the stratosphere absorbs nearly all of the biologically damaging solar ultraviolet radiation before it reaches the Earth's surface where it can cause skin cancer, cataracts, and immune deficiencies, and can harm crops and aquatic ecosystems. See ozone layer.

Ozone is produced naturally in the middle and upper stratosphere through dissociation of molecular oxygen by sunlight. In the absence of chemical species produced by human activity, a number of competing chemical reactions among naturally-occurring species—primarily atomic oxygen, molecular oxygen, and oxides of hydrogen and nitrogen—maintains the proper ozone balance.

In the present-day stratosphere, this natural balance has been altered, particularly by the introduction of man-made chloroflourocarbons. If the ozone decreases, the ultraviolet radiation at the Earth's surface will increase. See greenhouse gas.

Tropospheric ozone is a by-product of the photochemical (light-induced) processes associated with air pollution. Ozone in the troposphere can damage plants and humans.

Ozone hole A large area of intense stratospheric ozone depletion over the Antarctic continent that typically occurs annually between late August and early October; and generally ends in mid November. This severe ozone thinning has increased conspicuously since the late seventies and early eighties. This phenomenon is the result of chemical mechanisms initiated by man-made chlorofluorocarbons (see CFCs). Continued buildup of CFCs is expected to lead to additional ozone loss worldwide.

The thinning is focused in the Antarctic because of particular meteorological conditions there. During Austral spring (September and October in the Southern Hemisphere) a belt of stratospheric winds encircles Antarctica essentially isolating the cold stratospheric air there from the warmer air of the middle latitudes. The frigid air permits the formation of ice clouds that facilitate chemical interactions among nitrogen, hydrogen, and chlorine (elevated from CFCs) atoms, the end product of which is the destruction of ozone.

Ozone Layer The layer of ozone that begins approximately 15 km above Earth and thins to an almost negligible amount at about 50 km, shields the Earth from harmful ultraviolet radiation from the sun. The highest natural concentration of ozone (approximately 10 parts per million by volume) occurs in the stratosphere at approximately 25 km above Earth. The stratospheric ozone concentration changes throughout the year as stratospheric circulation changes with the seasons. Natural events such as volcanoes and solar flares can produce changes in ozone concentration, but man-made changes are of the greatest concern.

Ozone measuring satellite instruments Satellite-based ozone-measuring instruments can measure ozone by looking at the amount of ultraviolet absorption reflected from the Earth's surface and clouds. Some instruments provide data within the different levels of the atmosphere. The Total Ozone Mapping Spectrometer (TOMS) maps the total amount of ozone between ground and the top of the atmosphere.

The amount and distribution of ozone molecules in the stratosphere varies greatly over the globe, changing in response to natural cycles such as seasons, sun cycles. and winds. Utilizing satellites has enabled scientists to assess ozone levels simultaneously over the entire Earth, and has led them to conclude that global ozone levels are being depleted.

Ozone mini-hole(s) Rapid, transient, polar-ozone depletion. These depletions. which take place over a 50-kilometer squared area, are caused by weather patterns in the upper troposphere. The decrease in ozone during a mini-hole event is caused by transport, with no chemical depletion of ozone. However; the cold stratospheric temperatures associated with weather systems can cause clouds to form that can lead to the conversion of chlorine compound from inert to reactive forms. These chlorine compounds can then produce longer-term ozone reductions after the mini-hole has passed.

P

Palatability The characteristic of plants or plant parts that elicit a selective grazing or browsing response by animals. "Palatability" is controlled by the plant factors of chemical composition, proportion of plant parts, growth stage, external form of plant parts; influenced by environmental factors such as slope steepness, wind, sun or shade; and the animal factors of instinct, learning, physiological state, individual behavioural pattern variations and animal sensory responses.

Paleoclimate Climate as it existed in the distant past, particularly before historical records.

Paleogeography The study of ancient or prehistoric geography.

Paleontological Area Areas which have been designated as containing significant remains (usually fossilized) of flora and fauna (nonhuman) of geologic time periods before the appearance of humans.

Palustrine Vegetated wetlands dominated by trees, shrubs, and persistent emergents. It typically includes areas traditionally called marsh, swamp, bog, fen, prairie, and also includes small shallow, permanent or intermittent water bodies called ponds less than 6.6 feet deep. These wetlands, except for ponds, are situated shoreward of lakes, river channels, and large river embayments.

Panchromatic Sensitive to all or most of the visible spectrum.

Paper Plan (Map Plan) A term sometimes used to describe formal plan documents which consist mainly of a map on which land use regulation districts are indicated. While such explanatory written information as is necessary to convey its meaning accompanies a "paper plan", the basic regulatory document is the map. In contrast, plans whose basic regulatory document consists of a set of policies are sometimes called policy plans, whether accompanied or not by maps of the planning.

Paradigm A collection of the major assumptions, concepts, and propositions in a substantive area of work or knowledge, a coherent body of ideas or an approach (though not a technique) attracting widespread acceptance and use and providing the conceptual and contextual status for a branch of knowledge. Paradigms serve to orient research and theorizing in an area, and resemble models. Paradigms describe "approaches" and compactly codify central concepts and their relations.

Paradigmatic Tending to represent a paradigm. Tending to take a particular, nominal approach to a profession, discipline, or problem.

Parameter 1. A characteristic element or constant factor, such as an agency code, streamflow, aquifer type, etc.;.

2. A word used as a symbol for a variable;.

3. The name for a quantifiable variable or factor.

Parameter Code A code established to identify a specific type of data collected, and stored in a database. An example is the eight-digit code that identifies the type of water-related data available from an organization.

Parameter, Biological 1. A quantitative measure that characterizes or describes a statistical population (e.g., a population height); estimated by a sample statistic.

2. Systems ecology context – a quantified estimate or measurement of the value of an attribute of a component of an ecological system; e.g., 60 g/m is a measure of biomass for some species or group of species of organisms for a given site.

Parametric String A predefined command containing one or more variables for which the user supplies values at the time a computer command is invoked.

Parent Material The unconsolidated, chemically weathered mineral or inorganic matter from which the upper layers (i.e., where living roots are primarily found) of soil develops.

Parent Plan A plan establishing the decision framework in which other plans are made.

Parent Rock The original rocks from which sediments or soil were derived to form later rocks.

Pareto Optimum In welfare economics, a concept that sets a condition necessary to maximize the economic wealth of a given society. The pareto optimum is achieved when it is impossible to make one person better off without making another (or others) worse off.

Parity The addition of one or more redundant bits to information to verify its accuracy.

Part Per Million (PPM) One part by weight of dissolved chemical, or suspended sediment, in 1 million parts by weights of water.

Partial Retention A visual quality objective (VQO) in which human activities may be evident but must remain subordinate to the characteristic landscape.

Particle Size The general dimensions (such as average diameter or volume) of the particles in a sediment or rock, or the grains of a particular mineral that make up a sediment or rock. It is usually measured by sieving, by calculating settling velocities, or by determining areas on an image.

Pascal (Pa) Unit of atmospheric pressure named in honour of Blaise Pascal, whose experiments greatly increased knowledge of the atmosphere. A Pascal is the force of one Newton acting on a surface area of one square meter. It is the unit of pressure designated by the International System.

Passive system A system sensing only radiation emitted by the object being viewed or reflected by the object from a source other than the system.

Pasture An area devoted to producing forage (introduced or native); grass or other growing plants harvested by grazing and usually enclosed and separated from other areas by a fence or wall.

Pasture Management The application of practices to keep pasture plants growing actively over as long a period as possible so that they will provide palatable feed of high nutritive value; to encourage the growth of desirable grasses and legumes while crowding out weeds, brush, and inferior grasses.

Pasture, Annual A pasture consisting of introduced forage species planted for only one year or season's duration.

Pasture, Native Land used for grazing on which the climax (natural potential) plant community is forest but which is used and managed primarily to produce native or naturalized plants for forage. Includes cutover forestland, pastureland, or cropland originally cleared from forest and managed for native or naturalized forage plants.

Pasture, Permanent A pasture (usually created) consisting mainly of introduced perennial plants which are permitted to remain for a number of years.

Pasture, Rotation A pasture of introduced perennial and/or self-seeding annual species, comprising a unit in a rotation (deferred, rest, etc.) of 5 years duration or less.

Pastureland Land used and managed primarily for producing adapted domesticated forage plants to be grazed by livestock. Management practices usually include one or more treatments such as reseeding, renovating, mowing, liming, or fertilizing. Native pasture, because of location or soil limitation, may be treated like rangeland or included in this concept.

Patch Cutting A clearcutting system variation in which the logging operation removes all merchantable timber in patches of about 10 to 100 acres, separated for as long as practicable (preferably until the regeneration is adequately shading the forest floor) by living forest, so as to secure both the optimum dispersal of seed and to avoid the high hazard of large continuous areas of logging debris. "Patch cuts" are logging operations of a size generally which does not include the entire stand of which they are a part.

Pathogen A living organism capable of causing disease in a particular species or range of species.

Pathological Rotation The maximum age for a stand when reduced tree vigor occurs and before major insect and disease losses typically occur.

Pathology, Forest The science that deals with diseases of forest trees or stands, and with the deterioration of forest products by organisms.

Payments to Towns The money that is paid to local governments resulting from the presence of management activities on central or state lands within a town boundary.

Peak Discharge The maximum volume of flow attained at a given point in a stream during a runoff event.

Peeler A log from which veneer stock may he cut.

Perch The length of a rod, or a pole, i.e., 16.5 feet or 1/4th of a chain (old survey records) ; also the name of a game fish.

Percolation The downward movement of water within a soil, especially the downward flow of water in saturated or nearly saturated soil. Infiltration connotes entrance and exit of a substance through a soil.

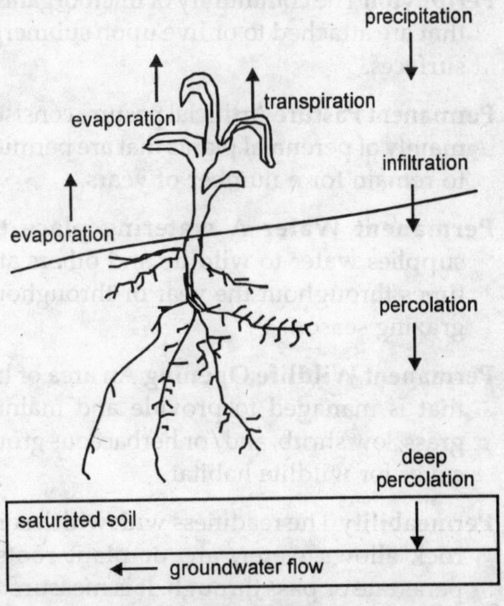

Fig. Percolation

Perennial Stream Streams that flow throughout the year from source to mouth and are typically shown in solid blue on minute quadrangle maps.

Perennials Plants that form annual above ground vegetation and seed structures from underground roots that persist for many years.

Performance Measure Indicators used to quantify outcomes of desired conditions, goals, objectives, or outcomes.

Perihelion The point in the orbit of a planet or comet which is nearest the Sun (as opposed to the aphelion, which is the point in the orbit farthest from the Sun).

Period Time required for a satellite to make one complete orbit.

Period decay (decay) The tendency of a satellite to lose orbital velocity due to the influence of atmospheric drag and gravitational forces. A decaying object eventually impacts the surface of the Earth or burns up in the atmosphere. This parameter directly affects the satellite's mean motion.

Periphyton The community of microorganisms that are attached to or live upon submerged surfaces.

Permanent Pasture Artificial pasture consisting mainly of perennial plants that are permitted to remain for a number of years.

Permanent Water A watering place that supplies water to wildlife and others at all times throughout the year or throughout a grazing season.

Permanent Wildlife Opening An area of land that is managed to provide and maintain grass, low shrub, and/or herbaceous ground cover for wildlife habitat.

Permeability The readiness with which soil or rock allows water, air, or plant roots to penetrate or pass through. It is measured by the rate at which water can enter and move through soil in a given interval of time under standard conditions.

Permeability, Coefficient of An expression of the rate of flow of a fluid through a unit cross section of porous mass under a unit hydraulic gradient, at a temperature of 60 degrees Fahrenheit. The standard coefficient of permeability used in the hydrologic work is defined as the rate of flow of water at 60 degrees Fahrenheit, in gallons or million gallons a day, through a cross section of 1 sq. ft., under a hydraulic gradient of 100%.

Persistence Time The time required for a pesticide to become inert. Arbitrarily assumed to equal four half-lives when measured persistence time is not available.

Persistent Pesticides Pesticides composed of compounds that retain their toxicity for long periods after application, e.g., DDT and other chlorinated hydrocarbons. Such pesticides do not deteriorate quickly, work their way up the food chain through various animals, and may accumulate in the fatty tissues of animals, including people, and remain there.

Personal Space An area with invisible boundaries surrounding a person's body into which intruders may not come without an effect. It is not necessarily spherical in shape. It does not extend equally in all directions. An invasion of "personal space" is an intrusion into another person's perceived boundaries of "self." Individual distance and personal space interact to affect the distribution of people.

Personnel People (or the count of such people) employed by an agency or land owner whose salaries or wages are paid from appropriated, allocated, or trust funds. It typically does not include contractors, cooperators, volunteers, or trainees.

Perspective An observation having the effects of distance upon the perceived size and characteristics of objects.

Perspective, Aerial The effects of distance from the viewer upon the colour and distinctness of objects – especially as due to the transparency of the intervening air. Typically objects become bluer, greyer, edges less distinct, and have reduced contrast of light and shade with increasing distance from the viewer.

Perspective, Linear The effects of distance from a viewer upon the appearance of the size and form of objects.

PERT Program Evaluation and Review Technique, also called "critical path method"; a planning aid, usually computerized.

Perturbation, Environmental Either an action or results of an action; a disturbance that occurs in an environmental system, which results in a change on components of that system. Perturbations can be classified as active or passive. Evolutions may reveal whether they are "good" or "bad." An active perturbation is the dynamic direct result of some activity; it is dynamic in that it moves through time or space. Emissions of any type that are injected into the environmental system are active; a smokestack is considered passive: it impedes the wind; it casts a shadow.

Perturbations Minor corrections to the Keplerian model of a satellite orbit as an ellipse of constant shape and orientation. Since satellite orbits are affected by Earth's gravity and drag caused by the Earth's atmosphere (causing satellites to spiral downward), minor adjustments must be made to the orbit.

Pesticide A general term used to refer to chemicals used to destroy, prevent, or control plant or animal pests. There are several classes of pesticides, defined by the type of pest they will influence, (e.g., herbicides, insecticides, rodenticides, molluskicides, and fungicides.).

PH A symbol for the degree of acidity or alkalinity of a solution. Expressed as a negative logarithm of the hydrogen ion concentration in a solution, $pH = -\log10[H^+]$. If the hydrogen ion concentration of a solution increases, the pH will decrease, and vice versa. The value for pure distilled water is regarded as neutral, pH values from 0 to 7 indicate acidity, and from 7 to 14 indicate alkalinity.

Phase interval In direct readout, the time between the end of a satellite image start tone and the start of the actual frame data. The phase interval repre sents white level video, interrupted by a black level pulse marking the start of each line and is used to set up phasing prior to image display.

Phenology Subdiscipline of agriculture, a science that treats relations between climate and periodic biological phenomena that are related to or caused by climatic conditions, such as the budding of trees and the migration of birds.

Phloem The tissue in higher plants which transports organic nutrients manufactured in the leaves to other portions of the plant.

Photochemical smog A type of smog that forms in large cities when chemical reactions take place in the presence of sunlight, its principal component is ozone. Ozone and other oxidants are not emitted into the air directly but form from reactions involving nitrogen oxides and hydrocarbons. Because of its smog-making ability, ozone in the lower atmosphere (troposphere) is often referred to as "bad" ozone.

Photomontage A combination of several distinct photographic pictures so they often blend with or into each other to produce a composite picture. Also contrasting shots or sequences may be blended for the purpose of suggesting a total idea or impression, or developing a theme.

Photon A quantum (smallest unit in which waves may be emitted or absorbed) of light.

Photoperiod The duration of light during a 24-hour period.

Photosynthetically active radiation Electromagnetic radiation in the part of the spectrum used by plants for photosynthesis.

Phreatophyte A type of plant that habitually obtains its water supply from the zone of groundwater saturation, either directly from within the ground water level or from the capillary fringe just above it.

Phyllosphere The interactive biological, physical, and chemical environment of plant leaf surfaces. These surfaces may have specialized biofilms and are invaded by bacteria, fungi, other microbes as well as insects.

Physical climate system The system of processes that regulate climate, including atmospheric and ocean circulation, evaporation, and precipitation.

Physical Life That period of time that a facility exists, not necessarily in a useable condition or used. Often facilities are removed from active service and kept intact some years before they are junked or destroyed, or they may be just left in their service position but never used.

Physical Planning A form of urban land use planning which attempts to achieve a physically attractive urban open space environment by regulating the physical relationships of structures to their sites and surroundings – with such devices as set back requirements, building height controls, maximum lot coverage by structures and unnatural surfaces, etc. The intent of such physical standards is to avoid claustrophobic and ugly urban open space environments. They may result in monotonous cities.

Physiographic (Morphologic) Descriptive of the surface form and features of the Earth, as bodies of air, water and land; equivalent to geomorphologic. Geomorphology is a study of landform genesis whereas "physiography" may be only descriptive.

Physiographic Determinism A theory that the natural elements of an environment should be respected in planning and given priority consideration in arriving at policy determinations. For example, "nature" should have priority in planning "aquifers" and wetlands, and the rest of nature's assets should be preserved in any development plan. It is equivalent to ecological determinism and biophysical determinism. "Physiographic" only refers to the nonliving elements of an environment while "physiographic determinism" encompasses the living elements as well.

Physiographic Production Controls over the amount occurring or potential biotic productivity which are due to abiotic environmental factors such as regional climate, geologic materials, slope orientation, and elevation.

Physiographic Site Class Units created by isolating areas within land-types on the basis of local climate considerations – such as slope orientation, air temperature and humidity. Variation in plant communities is used as the principal indicator of climate variation.

Physiographic Site Phase Created by adding information on the use requirements (e.g., beach slope and width, water depth, ridge tops, valley bottoms, toe-slopes, etc.) to a map of "physiographic site types". This is the smallest mappable unit land use or ecological unit. The various levels of G.A. Hills' categories in order of decreasing size, are site region, land-type, physiographic site class, physiographic site type, and physiographic site phase.

Phytoplankton Suspended or floating plant organisms, such as diatoms and blue-green algae, which drift passively with water currents.

Phytotoxicity Injury to plants due to exposure to a chemical.

Pickaroon A device with a head similar to an axe but with a point rather thian a blade mounted on tie end of a handle which is used to assist in lifting and placing bolts of wood.

Piling Round timbers to be driven into the ground to support other structures.

Pine Conversion Establishing pines on sites that were formerly occupied by hardwoods to yield trees of high quality and volume per acre.

Pitting Making shallow pits or basins of adequate capacity and distribution on range to speed range improvement by retaining water from rainfall and snowmelt and by reducing competition to desirable species from a dense stand of less desirable vegetation.

Pixel Smallest part (addressable element) of an electronically-coded image, such as a computer display. Pixel is a contraction of "picture element.".

Placer Claim A mining claim located upon a mass of gravel, sand, or similar material (resulting from weathering, erosion and/or stream deposition) containing particles of gold, platinum, tin, or other valuable minerals and whose valuable mineral contents are extracted by concentration in running water. The maximum size of an individual "placer claim" is 600 by 1320 feet (10 acres).

Plan Element The major land use planning elements such as legal requirements and authorities, national and regional objectives and targets, management situations, basic assumptions, data collection, land capability, alternative considerations, plan selection, functional planning and documentation. Also a discrete topic area in an overall plan, e.g., housing, open space, recreation, transportation.

Plan Selection Selection of one plan from a set of alternative plans. Subjected to the planning process, a choice is made governed by a reasonable and rational perception of how the available resources can be most efficiently managed to achieve optimum multiple benefits for the people or a landowner. The process includes full consideration of the priorities and preferences expressed by the public or the responsible landowner at all levels to be affected by the plan.

Planetary albedo The fraction of incident solar radiation that is reflected by a planet and returned to space. The planetary albedo of the Earth-atmosphere system is approximately 30 percent, most of which is due to backscatter from clouds in the atmosphere.

Plankton Small (usually less than 2mm long) floating or drifting life forms in water bodies. Plankton includes both plants (phytoplankton) and animals (zooplankton) that are carried passively in the water currents. Those that can swim do so to change or adjust their depth in the water, not to move from place to place. Plankton is one of the 3 main divisions of aquatic life. The others being "nekton" (the animals that swim actively and may move long distances for feeding or breeding) and the "benthos" (organisms which crawl about on the bottom, or burrow into the bottom or grow attached to the bottom).

Planned Grazing System A system of grazing in which two or more grazing units are alternately rested from grazing in a planned sequence over a period of years. Resting the land may be throughout the year or during the growing season of the key plant species.

Planned Ignition A fire started by a deliberate management action.

Planning 1. Creating plans. A dynamic effort to use decisions to guide future actions and decisions, one element in the process of guiding the future.

2. Determining the objectives or goals of an enterprise and the selection, through a systematic consideration of alternatives, of the policies, programs, procedures, and schedules for achieving them reporting, and adapting to achieve effectiveness.

3. Designing a future organization, structure, and relationship of parts.

4. The process of finding alternative solutions to a problem. A way of thinking about and creating solutions for future problem.

5. A nondescript process oriented predominantly toward the future and deeply concerned with the relation of objectives to collective decisions in policy and program.

6. A process of conflict resolution (usually future) by designing optimum relationships between components without attempting to favour or eliminate any party in the (future) conflict.

7. With emphasis on material and inventory, an effort to attain some future objectives by rationally "determining" in the present the requisite resources, facilities, services, personnel, and procedures (and their future availability) needed to achieve those objectives.

8. Preparing a land use zoning map.

9. Preparing a schedule of activities.

10. Thinking about the future.

Planning Agent The person(s) or organization actually doing the planning. A "planning agent" may be a person (an individual householder or entrepreneur), a family, a legal person (a corporation, a trade association), or various levels of government (central, state, county, public district).

Planning Area A specifically identified geographic area containing social and physical resources and land characteristics of a generally similar nature. Because general similarities in physical characteristics, resources, economics, and social needs give rise to similar management opportunities and constraints, therefore planning efforts, irrespective of administration or government boundaries, can often be coordinated. A variety of considerations may interact to determine the size and boundaries of a planning area; topography, vegetation or broad habitat type, climate, history of land use, populations – human and animal, economy and the relationship of these and other elements to each other will affect the definition of a planning area. These areas may, and will likely, include several administrative boundaries. A planning area should be of sufficient size to allow for flexible patterns of land to develop and resources to be used. State and multi-county planning districts, and the integrity of counties, are strong considerations in establishing planning area boundaries. Commonality of boundaries will aid in cooperative planning and data exchange. The area should permit a maximum of agencies or governments to contribute and receive resource data and relate demand allocations with high compatibility. Coordinated demand projections for the resources within a land area can be efficiently gained in large planning areas.

Planning Criteria Standards, terms, tests, rules, and guidelines by which the land and resource management planning process is conducted, and upon which judgments and decisions about the plan are based.

Planning Horizon The overall period considered in the planning process that spans all activities covered in the analysis or plan, and all future conditions and effects of proposed actions which would influence the planning decisions. For timber resource planning purposes, the overall time period extends far enough into the future to assure that the allowable sale quantity can be achieved and maintained at the long-term sustained yield capacity. It is the last year of the planning period (fixed by the planner), the length of the planning period that will be considered in the planning process.

Planning Level The scale of a planning effort, usually the size of the project or the governmental level at which the planning is being done, e.g., site or project planning, county planning, state planning, interstate regional planning.

Planning Period Usually one 10-year period. The time interval within the planning horizon that is used to show incremental changes in yields, costs, effects, and benefits.

Planning Problem A major problem of long-range significance, derived from public issues and management concerns, to be addressed when formulating plans.

Planning Unit Geographic areas of land, varying in size, that are characterized by particular patterns of topography, climate, and land use. A unit may occur as one major drainage, several separate drainages, or portions of drainages. These are specific, identified, and delineated land areas large enough to encompass most of the meaningful or significant relationships within a drainage or series of drainages. Topographic features are logical places for planning unit boundaries because of the need to be able to identify boundaries on maps and photographs, and on the ground. Many factors influence the boundaries, size, and layout of planning units, among which are: 1. intensity of planning expected;.

2. The resources available for doing planning, people and time;.

3. Presence or absence of unique features;.

4. perceived complexity or homogeneity of the land features;.

5. Public involvement inputs; and.

6. Planning team judgment. The objective is to preserve as much uniformity as possible in relationships significant to land management. Such factors as existing transportation systems, population influences, current land use, topography, soils, industrial dependency, ownership, and public feelings are considered in establishing the boundaries for planning units. No two planning units are exactly equivalent in the factors that influence the boundaries of the planning unit. The purpose of planning units is to organize planning activities and concentrate them on areas small enough to be workable and large enough to enable a planning team to envision or predict the cause and-effect relationships of management alternatives. No size limits are arbitrarily established for planning units. Planning unit boundaries may cross both planning area and administrative unit boundaries.

Planning, Anticipatory Any planning approach which operates by attempting to foresee potential issues or problems and develop solutions to them before they become real. An aspect of feedforward.

Planning-Programming-Budgeting System (PPBS) A planning and decision-making tool that attempted to organize information and analysis so that the consequences of alternative policies are clearly revealed and fully comparable. Under PPBS (ordered instituted in each Cabinet agency by President Johnson in 1966), programs having the same purpose are grouped together under the same category and compared to see which represents the best use of the government's money. PPBS endeavors to display the allocation of resources in terms of the function; to gather quantitative information on the actual results of programs; and to analyze the cost of the various efforts to achieve similar ends and rank them in terms of their effectiveness per dollar.

Plant Community Any named aggregation of plants having high uniformity in composition and structure and occupying an area of essentially uniform environment. Each of these communities is dominated by certain species which are conspicuous or which are rare. Even when areas of equivalent environment are encountered, whether continuous or detached, essentially the same plant assemblage reappears and there is a pattern to their appearance.

Plant Retrogression The process of deterioration in vegetation whereby an area becomes successively occupied by different plant communities of lower (earlier) ecological order in succession.

Plant Species, Decreaser Plant species of original range vegetation that will decrease in relative amount in population density or cover with continued grazing or overuse.

Plant Species, Increaser Plant species of the original vegetation that increase in relative amount, at least for a time, under overuse.

Plantation A forest crop or stand raised artificially, either by seeding or planting of young trees.

Plantation An artificially reforested area established by planting or direct, seeding. Contrast with a natural forest stand which is established naturally.

Planting Bar A hand tool used in making a slit-hole in which trees are planted.

Planting Levels Five tree-improvement program intensity levels ranked by increasing complexity based on workloads and potential genetic gain that can be realized.

Plasma A fourth state of matter (in addition to solid, liquid, and gas) that exists in space. In this state, atoms are positively charged and share space with free negatively-charged electrons. Plasma can conduct electricity and interact strongly with electric and magnetic fields. The solar wind is actually hot plasma blowing from the sun.

Plat A scale drawing of essential data pertaining to boundaries, subdivisions, and ownership(s) of a tract of land, as determined by survey.

Plate tectonics Concept that the Earth's crust is composed of rigid plates that move over a less rigid interior.

Platforms A satellite that can carry instruments. The same term is applied to automatic weather data transmitters installed on buoys, balloons, ships, and planes, and mounted in remote areas.

Plot An area of land usually less than 1 acre on which trees and sometimes other vegetation are measured during a cruise (or inventory).

Pluralism A description of society that contends that political power is distributed among a variety of groups and institutions. Pluralist theory argues that political resources — wealth, education, prestige, skill, and votes — are divided and that no group has a monopoly over all resources. In contrast, elitist theories of power argue that small groups of individuals rule a country in their own economic interest. Pluralists agree that only small numbers of citizens participate directly in decision-making, but they argue that these elites compete for economic rewards, differ on policy, are open for others to join, and are indirectly influenced by voters. "Elitists" contend that members of the ruling power-elite share a pro-capitalist ideology, are unified by a similar upper-class background, and are linked by institution and personalities.

Pluralistic Society (Cultural Pluralism) A society in which there are widely differing and often conflicting points of view as to how the access to or use of a particular resource (or resources in general) should be allocated.

Poacher Any person who illegally takes for personal or group benefits a wild animal or wild plant. The act or acts related to purposefully attempting to take or actually taking a wild animal or plant of improper species, sex, time of year, excessive numbers, or area classify any person of any description as a poacher. People seem to be opportunistic and given a special situation may act as described above, thus becoming a poacher for the first time. Once a person has poached, there are (2005) no categories for those who will never do so agin, those who have stopped, or who are full-time engaged in the acts. Some people include all people engaged in all aspects of taking, processing, and selling wild animals and plants or their parts as poachers.

Point-Source Pollution (Point Pollution Source) Pollution the source of which is specific rather than general in location. Particulate matter emanating from a specific smokestack or toxic effluent from a pipe is point source pollution.

Polar orbit An orbit with an orbital inclination of near 90° where the satellite ground track will cross both polar regions once during each orbit. The term is used to desribe the near-polar orbits of spacecraft such as the USA's NOAA/TIROS and LANDSAT satellites.

Pole and Pole Timber A young tree 4 inches or more in diameter breast high.1.As used in timber survey, a size class definition, trees 5.0 to 8.9 inches in DBH. The maximum size of poles is usually though not invariably taken to be some diameter breast high between 8 and 12 inches.

2. As used in logging operations, trees from which pole products are produced, such as telephone poles, pilings, etc.

Police Power A state's inherent and plenary power over persons and property that enables the society to prohibit or regulate all things inimical to its comfort, safety, health, and welfare.

Policy A decided course or method of action selected by a governmental agency, institution, group or individual. It is selected from among alternatives and, in the light of given conditions, to guide and usually determine present and future decisions and actions. It is a guiding principle, typically a set of major constraints, influencing a specific decision or set of decisions.

Policy Issue An action or set of circumstances that has bearing on current or future policy, resulting in proposed policy, alternative policies, or an identified need for a policy.

Policy Plan A formal plan document consisting entirely of a set of policies unaccompanied by maps differentiating the planning area where some policies apply while others do not. In contrast, a plan whose formal plan documents consist mainly of a map on which land use regulation districts are indicated is sometimes called a paper plan or "map plan". While "paper plans" are accompanied by explanatory written information necessary to convey their meaning, the basic regulatory document is still the map.

Political Boundary A boundary which marks the limits of a political jurisdiction. The boundaries of cities, counties, states and nations are political boundaries. Lesser political boundaries limit tax districts, legislative districts, etc.

Political Determinism A particular point of view in planning which advocates that all resource use and allocation decisions should be made via the political process – i.e., by elected representatives and legislation. In theory, politics is society's means for attaining some needed resolution of issues via compromise from the often-conflicting demands on a resource by different groups. This point of view is that elected representatives' decisions are responsive to and representative of the collective judgments of their constituents. Politics thus is viewed as the process by which society makes group choices about resource use allocation, passes final judgment upon interim administrative decisions, and conveys the right (via legislation) to regulate resource use or the exercise of private property rights. It is believed that because political decisions are societal choices they are the best choices that can be made and thus should be the basis for making planning decisions.

Political Institutions A society's system of legal rules, principles, and organizations.

Pollution 1. Concentrations of substances or radiation in excess of those for which there has been a collective genetic experience by communities.

2. Any substance or energy form (heat, light, noise, etc.) which alters the state of the environment from what would naturally occur. Especially associated with those altered states which human value judgements have decreed as bad.

3. An activity of people, which directly or indirectly results in adding to water, air, or soil, matter or energy which has a deleterious effect on living organisms or structures it is desirable to preserve, or which reduces the quality of water, air, or soil for any subsequent use. In this context, such terms as natural pollution and visual pollution (not referring to decreased visibility due to air pollution) are meaningless and should not be used.

4. The presence of matter or energy (usually undesirable) in an unusual or unintended place.

5. The presence in the environment of one or more contaminants or nuisances in sufficient quantities and of such characteristics and duration as to be injurious or irritating to human, plant, or animal life, to health, or to property, or to unreasonably interfere with production or with the enjoyment of life or property.

Pond A small impoundment of water, usually 0.5 to 4.9 acres in surface area, usually with fish.

Population An aggregation of individuals of the same species inhabiting a given area at a time.

Population Dynamics 1. The totality of changes that take place during the life of a population.

2. A study of changes, and the reasons for those changes, in the numbers of individuals in populations.

3. A study of the changes among sub-populations over a landscape.

Population Pressure The combined effects exerted by the individuals of a population upon the other organisms in a community and upon their physical environment.

Population/Habitat Unit A discrete association of individual animal bonded together by traditional use of a habitat.

Porosity, Effective The ratio, usually expressed in percentage of (a) the volume of water or other liquid which a given volume of rock or soil, after being saturated with the liquid, will yield under any specified hydraulic conditions, to (b) the total volume of soil or rock.

Post A short timber up to 16 feet in length used in an upright position to support other structures for fencing.

Post Market The market of trees to be used as fence posts. They are normally 4 inches to 10 inches in diameter and 6 to 10 feet long.

Potable Water Water suitable for human drinking.

Potential Wilderness Area An area including those previously identified in the Forest Service Roadless Area Conservation, Final Environmental Impact Statement, Volume 2, dated November 2000, in a unit plan, or in a land management plan, which remain essentially roadless and undeveloped, and which have not yet been designated as wilderness or for non-wilderness uses by law.

Potential Yield The maximum yield of timber that can be obtained annually on a sustained yield basis by utilizing intensive forestry management practices.

Poverty, Absolute A standard of living below the minimum needed for maintenance of life and health.

Poverty, Relative A condition characterized by a scarcity rather than a lack of economic necessities. A relatively poverty-stricken person is poor in comparison with a majority of others in his society, but his or her level of living usually does not constitute an immediate threat to life or health.

Practical Maximum Capacity The maximum potential recreation visitor days produced by the recreation opportunity spectrum.

Prairie An extensive tract of level or rolling land, usually with deep fertile soil, that was originally treeless and grass covered.

Precession The comparatively slow torquing of the orbital planes of all satellites with respect to the Earth's axis, due to the bulge of the Earth at the equator which distorts the Earth's gravitational field. Precession is manifest by the slow rotation of the line of nodes of the orbit (westward for inclinations less than 90° and eastward for inclinations greater than 90°).

Precipitation Moisture that falls from clouds. Although clouds appear to float in the sky they are always falling, their water droplets slowly being pulled down by gravity. Because their water droplets are so small and light, it can take 21 days to fall 1,000 feet and wind currents can easily interrupt their descent. Liquid water falls as rain or drizzle. All raindrops form around particles of salt or dust. (Some of this dust comes from tiny meteorites and even the tails of comets.) Water or ice droplets stick to these particles, then the drops attract more water and continue getting bigger until they are large enough to fall out of the cloud. Drizzle drops are smaller than raindrops. In many clouds, raindrops actually begin as tiny ice crystals that form when part or all of a cloud is below freezing. As the ice crystals fall inside the cloud, they may collide with water droplets that freeze onto them. The ice crystals continue to grow larger; until large enough to fall from the cloud. They pass through warm air; melt, and fall as raindrops.

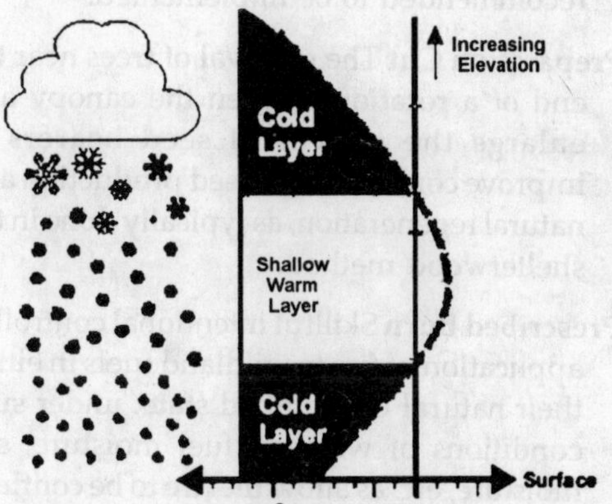

Fig. Precipitation

When ice crystals move within a very cold cloud (10°F and -40°F) and enough water droplets freeze onto the ice crystals, snow will fall from the cloud. If the surface temperature is colder than 32°F the flakes will land as snow.

Precocial Young born with eyes open, down or fur covered and quite mobile in the first day or two of life, e.g. ruffed grouse young.

Pre-Commercial Operations Cutting conducted in forest stands which removes wood of a size too small to be marketed. Such operations usually are designed to improve species composition and increase quality, growth, and vigor of the remaining trees.

Pre-commercial Thinning Reduction of the number of trees to improve the growth rate and quality of the remaining trees on areas where the trees that are cut do not have an economic value or market.

Predator A bird or mammal species that feeds upon live animals; a carnivore.

Preferable Species Plant species that are preferred by animals and are grazed during first choice.

Preferred Alternative The alternative recommended to be implemented.

Preparation Cut The removal of trees near the end of a rotation to open the canopy and enlarge the crowns of seed bearers to improve conditions for seed production and natural regeneration, as typically done in the shelterwood method.

Prescribed Burn Skillful intentional controlled application of fire to wildland fuels in either their natural or modified state, under such conditions of weather, fuel moisture, soil moisture, etc. as allow the fire to be confined to a predetermined area and to produce the intensity of heat and rate of spread required to further certain planned objectives of silviculture, wildlife management, grazing, fire-hazard reduction, etc. It seeks to employ fire so as to realize maximum net benefits with minimum damage and at acceptable cost. The fire may result from either planned or unplanned ignitions.

Prescribed Fire Any wildland or rural fire ignited by management actions and burning under certain predetermined conditions to meet specific objectives such as those related to hazardous fuels, forest or range regeneration, or watershed or faunal space management. A planned fire used to dispose of fuels, control unwanted vegetation, restore or stimulate growth of desired vegetation, change successional stages, maintain fire dependent communities, reduce arthropod disease vector abundance, increase grazing wild faunal or livestock forage, and improve viewscapes in order to meet wildlife, recreation, wilderness, watershed, or timber management objectives.

Prescription A document for a timber compartment (or similar unit of land) which: 1. Provides data about the age, forest type, and conditions of the stands in the compartment;.

2. Determines the type of silvicultural and other treatment needed by the stands; and.

3. Specifies management that is needed to secure the desired benefits from the land over the long-run.

Present Net Value (PNV) The difference between the discounted aggregate value (benefits) of all outputs to which monetary values or established market prices are assigned and the total discounted costs of managing the area during a stated period and at a stated discount rate.

Present Worth The series of benefits or costs that occur with a program or project over time discounted back to present time. That amount of money a person would have to deposit now, at a given rate, to accumulate to a stated future value for *either* benefits or costs.

Preservation 1. A visual quality objective that allows for natural or ecological changes only.

2. Complete protection, no use, and minimum management.

Pressure Group (Interest Group, Special Interest Group) Any group of people with special interests which actively advocates for special consideration of its resource use goals. These are usually politically active groups with a common set of goals, values, or ideology about resource use allocation. They may influence decisions on resource use allocation in excess of their proportional representation in the planned-for populace by seeking preferential consideration for their resource use choices, such as lobbying, raising money for political campaigns, promising to "deliver" votes on election day or workers in political campaigns, developing some special expertise, or having some special relationship with an agency in the government bureaucracy. Pressure groups may be distinguished from the broader category of special interest groups by their somewhat more formal organization and their more noticeable activity over time. Interest groups are the most important unit of analysis in pluralist theory, for they mediate between an individual citizen's desires and the political parties. Two major differences between parties and interest groups are that parties compete for office, whereas interest groups do not, and interest groups present a homogeneous interest whereas parties have a heterogeneous clientele.

Pre-suppression Activities required in advance of fire occurrence to ensure an effective suppression action. Included are 1. Recruiting and training fire forces;.

2. Planning and organizing attack methods;.

3. Procuring and maintaining fire equipment; and.

4. Maintaining structural improvements necessary for the fire program.

Presuppression, Fire Activities in advance of fire occurrence to insure effective suppression action. Includes recruiting and training, planning the organization, maintaining fire equipment and fire control improvements, and procuring equipment and supplies.

Prevailing westerlies Winds in the middle latitudes (approximately 30° to 60°) that generally blow from west to east. The subtropical high pressure regions at the horse latitudes (30°) forces surface air poleward, and the rotation of the Earth causes these winds to bear to the right (east) in the Northern Hemisphere and to the left (east) in the Southern Hemisphere. This is, to some extent, an idealized picture of the atmospheric circulation. The actual circulation on individual days includes modifications and variations due to the migratory cyclones and anticyclones of middle latitudes, causing rapid and often violent weather changes, as warm semi-tropical air from the horse latitudes meets cold polar air from the high latitudes. See wind.

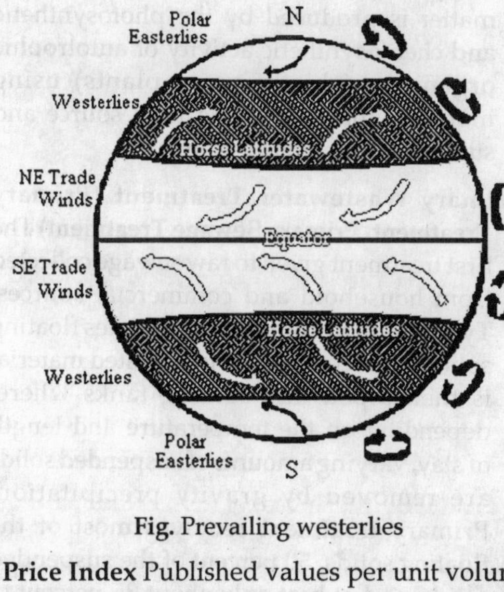

Fig. Prevailing westerlies

Price Index Published values per unit volume by tree species that are used to determine the actual price of cut timber. If the the index is lower at the time of cut that at the time of sale, the sale price is reduced by the decrease. If the index is higher, the sale price is increased by half of the increase.

Primary Conversion The initial steps in the process of converting a felled or fallen tree to lumber, e.g., topping, trimming limbs, removing bark, and sawing into logs.

Primary Facility Access and service to large land areas encompassed and served by a public transportation system. They include public transportation facilities and form the basic framework around which the transportation network is designed. A substantial percentage of the traffic served by these facilities is public service traffic, and emphasis is given to travel speed and efficiency.

Primary Output The main goods, services, or environmental conditions of a resource element - the key quantitative indicators for determining if objectives have been attained (e.g.,animal-unit-months of forage for livestock or board feet of timber harvested).

Primary Productivity The rate at which organic matter is produced by the photosynthetic and chemosynthetic activity of autotrophic organisms (chiefly green plants) using inorganic material as a carbon source and sunlight as an energy source.

Primary Wastewater Treatment (Primary Treatment, Primary Sewage Treatment) The first treatment given to raw sewage collected from household and commercial sources. Typically the process first separates floating solids by screening. The undigested material is then piped into settling tanks where, depending on the temperature and length of stay, varying amounts of suspended solids are removed by gravity precipitation. Primary treatment removes most of the floating solids, 50 percent of the suspended solids, and at best only about 35 percent of the organic waste and biochemical oxygen demand from wastewater.

Prime Agricultural Land Land identified (after considering the long-term productive potential and topographic position of the soil(s) climatic factors, and availability and quality of water) as being required for the economic production of specific agricultural commodities that are designated by local units.

Prime meridian An imaginary line running from north to south through Greenwich, England, used as the reference point for longitude.

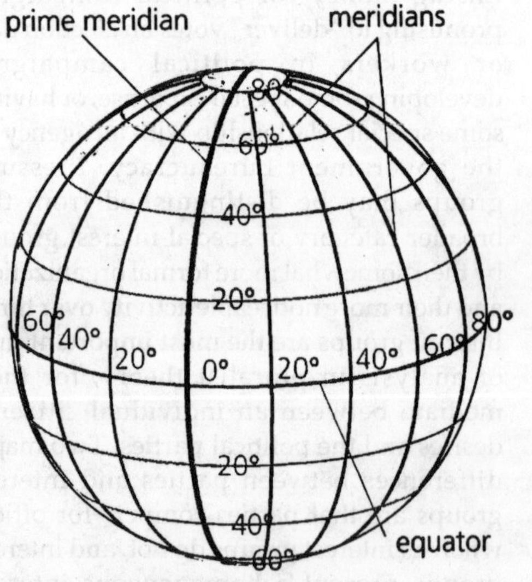

Fig. Prime meridian

Primitive Area 1. A large tract typically within Forest set aside for preservation in natural condition with no alteration or development beyond measures for fire protection being permitted. Within any area so designated no occupancy under the special-use permit shall be allowed, or the construction of permanent improvements by any public agency be permitted, except as authorized by Confounding instruction subsequently sent indicated that "the establishment of a Primitive Area ordinarily will not operate to withdraw timber, forage or water resources from industrial use, since the utilization of such resources, if properly regulated, will not be incompatible with the purposes for which the area is designated."

These areas are extensive natural, wild and undeveloped areas, usually in a setting far removed from the sights, sounds, and smells of civilization. Essential characteristics are that the natural environment has not been disturbed by commercial utilization and that the areas are without mechanized transportation. The area must be large enough and so located as to give the user the feeling that he or she is enjoying a wilderness experience. Activities in Primitive Areas involve camping out on one's own without mechanized transportation, permanent shelter, or other conveniences. People are only transitory visitor. There is no development of public roads, permanent habitations or recreation facilities except trails. No mechanized equipment is allowed except that needed for management and protection purposes. Some people use the terms "primitive" and "wilderness" interchangeably to refer to all areas displaying no obvious evidence of present or previous human uses-other than foot or pack animal trails. Others make a distinction between the two, using "primitive" to refer to all areas possessing wilderness qualities and restricting the use of "wilderness" to only those portions of the larger primitive area which have been designated as "Wilderness Areas.".

Primitive ROS Class The classification of the recreation opportunity spectrum characterized by an essentially unmodified environment, where trails may be present but structures are rare, and which probability of isolation from the sights and sounds of people are extremely high.

Printed circuit A fibre card on which integrated circuits and other electronic components can be mounted. Connections between the components are etched in the correct circuit patterns.

Private Property The exclusive right to the possession, use or disposal of property or goods. Private property can be real property, land, and the buildings fixed on it, or personal property that is movable property. The concept of "private property" is the creation of people in a society and thus property is subject to such intrusions upon it as society requires for the general welfare.

Process An association of phenomena governed by physical, chemical, or biological laws. An example of a process is the vertical mixing of ocean waters in the so-called surface-mixed layer; the state variables for this process include temperature, salinity in the water on a vertical scale of tens of meters, and heat flow and wind stress at the sea surface. Other examples include the volcanic deposition of dust and gases into the atmosphere, eddy formation in the atmosphere and oceans, and soil development.

Process study An organized, systematic investigation of a particular process designed to identify all of the state variables involved and to establish the relationships among them. Process studies yield numerical algorithms that connect the state variables and determine their rates of change; such algorithms are essential ingredients of Earth system models.

Producer An organism which can synthesize organic material using inorganic material and an external energy source (light or chemical).

Producer Organism (Producer, Autotrophic Organism) An organism, usually a green plant, that can utilize radiant energy (i.e., sunlight) to synthesize organic substances from inorganic materials.

Producer Surplus In economics, the difference between the price at which a producer sells a good or service and the amount that he or she would actually be willing to accept in exchange for the good or service.

Production Function A mathematical relationship (frequently presented as a graph) which shows how inputs of various physical factors are related to changes of output, e.g., the effect of different amounts of fertilizer or thinning on the rate of timber production. Expressions of wood or forage production over time are a special type of production function.

Productive Capacity of Ecosystems The continuing productivity of an ecological system, including its ability to sustain desirable conditions such as clean water, fertile soil, riparian habitat, and viable populations of plants and animals; and to sustain desirable human uses; and to renew itself following disturbance.

Productivity (Biological Context) 1. The rate of production of organic matter produced by biological activity in an area or volume (e.g., grams per square meter per day, or other units of weight or energy per area or volume and time).

2. The innate capacity of an environment to produce plant and animal life.

3. The capacity of a soil to produce a certain kind of crops under a defined set of management conditions.

4. The rate at which energy is stored by organisms in the form of organic substances which can be used as food materials.

5. The rate of storage of organic matter in tissue by organisms including that used by the organisms in maintaining themselves.

Productivity The capacity of National Forest System lands and their ecological systems to provide the various renewable resources in certain amounts in perpetuity. For the purposes of a Forest Service plan it is an ecological, not an economic, term.

Productivity, Biotic Growth products and by-products of living organisms (e.g., wood, meat, other plant fibres).

Productivity, Gross Primary The rate of synthesis of organic material produced by photosynthesis (or chemosynthesis), a rate that includes accumulating organic material in plant tissues.

Productivity, Net-Primary 1. The amount of organic material produced by biological activity in an area or volume.

2. The total amount of energy stored over a specified period by an organism or organisms.

Productivity, Primary (Basic Productivity) The rate at which energy is stored, typically as weight change, by the photosynthetic and chemosynthetic activity of producer organisms (chiefly green plants) in the form of organic substances.

Productivity, Secondary 1. The rate of new biomass addition (weight growth) and storage on plant-eating (herbivorous) animals.

2. The rate of energy storage by consumer organisms.

Productivity, Tertiary The rate of new biomass addition (weight growth) on meat-eating (i.e., carnivorous) animals.

Profile View An illustration in which the features coinciding in space with the plane of sectioning are strongly emphasized (usually by thick linework). Features in the background are usually shown with much thinner linework and without any reduction in size in proportion to their increasing distance from the plane of sectioning (as would occur if they were drawn according to the rules of linear perspective). Underground features are seldom shown in "profile" type drawings and there is never any exaggeration of the scale of vertical or horizontal features.

Prograde orbit Orbits of the Earth in the same direction as the rotation of the Earth (west-to-east).

Programmed Harvest The part of the potential yield that is planned for harvest in any one year (comparable to annual allowable harvest).

Programming The process of deciding on specific courses of action to be followed in carrying out planning decisions and objectives. It also involves decisions in terms of total inputs required or total costs to be incurred over a period of years as to personnel, material, and financial resources to be applied in carrying out programs.

Project An organized effort to achieve an objective identified by location, activities, outputs, effects, and time period and responsibilities for execution.

Project Design The process of developing specific information necessary to describe the location, timing, activities, outputs, effects, accountability, and control of a project.

Project Planning (Operational Planning) Types of planning concerned with physical undertakings (such as building a dam). Project planning incorporates a greater range of specific elements than functional planning and, thus, deals with more numerous and diverse parts.

Projection 1. An extrapolation of a series of data to estimate a value beyond the range of the data (usually using a linear or curvilinear model). These are estimates and may not have a time dimension.

2. Similar models may be used to predict future conditions and occurrences. The accuracy of a projection is very dependent on past trends continuing, or upon changing in some known manner. A "projection" differs in a strict sense from a forecast, which is also a prediction of future conditions and occurrences, but is based on an understanding of the system being discussed from current and past data rather than merely continuing and adjusting past trends. There is a distinction but it is imprecise.

3. Projections are conditional forecasts of the future. They are based upon stated assumptions about factors that are expected to influence future conditions. Projections reflect historical development and economic activity as modified by discernible trends determined by analysis of present and past relationships. If there are no changes in the objectives of the people, if limits of resource availability are not reached. If anticipated technology is attained, and if the influence of external forces remains constant, projections could reasonably be expected to chart the future.

Property The dominion or indefinite right of use or disposition which one may lawfully exercise over particular things or subjects, land, improvements or objects of value to which rights of ownership and control may be acquired. It includes the social relationships of rights and obligations between people over the use of an object. An object in these relationships may be material or nonmaterial, such as a trademark.

Property Rights The sum total of the rights and legal protection pertaining to property ownership. In the case of real property it embraces the right to quiet enjoyment, the right to sell, mortgage, lease, bequeath, regain possession at the end of a lease, build and remove improvements, and control the property's use within the law.

Prospecting Removing overburden, core drilling to obtain subsurface samples, constructing roads or any other disturbance of the surface for the purpose of determining the location, quantity or quality of a mineral deposit.

Prospectus A statement, which based on such items as budget, policies, and environmental changes, summarizes past performance and forecasts the course of key objectives. These anticipated levels of conditions, uses, and activities and related monitoring performance measures help describe progress towards attainment of desired conditions. A prospectus is usually identified within a plan's Strategy section.

Protection Activity that reduces losses or risks and that tends to maintain basic conditions and values and reduces damages or injury to potential or actual resource outputs of rural lands and resources. It includes insect and disease control, fire protection, and law enforcement. It may include teaching and other human behavioural-change activity.

Protective Element (Fire) The support element providing protection to forest resources and uses (such as fire suppression).

Proxemics The study of people's use of social and personal space.

Proxy Value A value assigned to a good or service for evaluation purposes when the good or service is not bought or sold and an established monetary price does not exist.

Public Involvement The opportunity for feasible participation by citizens in rulemaking, decision making, and land use planning, including public hearings, advisory mechanisms, and such other procedures. It includes informing the public, obtaining early and continuing public participation, and considering the views of interested parties in planning and decision-making processes. It encourages public understanding about and participation in the planning processes. The "public" includes individuals, local, state, regional and national public service organizations and interest groups.

Psychrometer An instrument designed to measure dew point and relative humidity consisting of two thermometers (one dry bulb and one wet bulb). The dew point and humidity levels are determined by drying the wet bulb (either by fanning or whirling the instrument) and comparing the difference between the wet and dry bulbs with preexisting calculations.

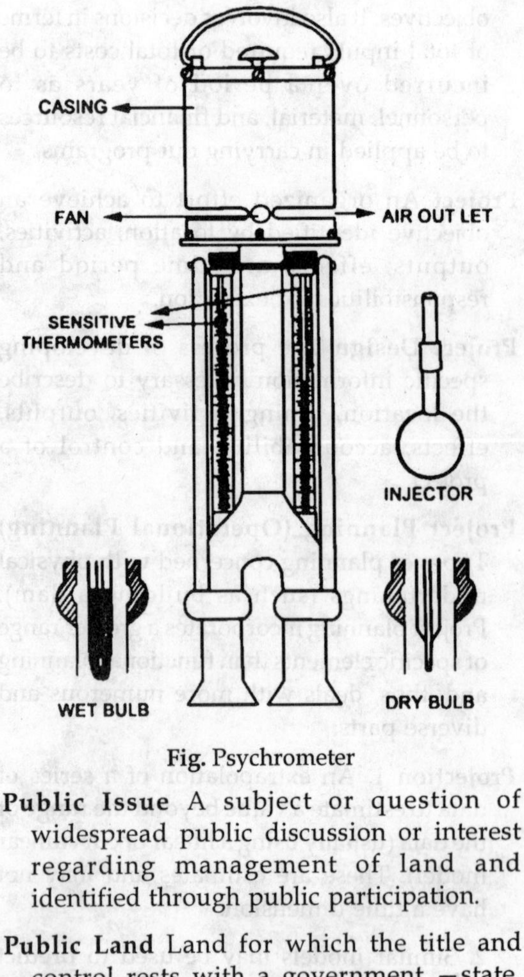

Fig. Psychrometer

Public Issue A subject or question of widespread public discussion or interest regarding management of land and identified through public participation.

Public Land Land for which the title and control rests with a government —state, regional, county, or municipal.

Public Participation Activities Meetings, conferences, seminars, workshops, tours, written comments, survey questionnaires, and similar activities designed or held to obtain comments from the general and specific publics.

Public Participation Activities that include a wide range of public involvement tools and processes, such as collaboration, public meetings, open houses, workshops, and comment periods.

Purchaser Credit Road Work Road work (construction or reconstruction) required of a timber purchaser in timber sale contracts, and paid for with timber receipts.

Put to Bed The practice of temporarily closing roads between periods of use. This involves removing all drainage structures, re-vegetating the roadbed and drainage ditches, and closing the area to vehicle traffic.

Pyramid of Biomass When the total weight of individuals at successive trophic levels in a food chain is plotted, a gradually sloping pyramid usually results, regardless of the individual size of the producer organisms or consumer organisms, and regardless of whether predators or parasites are considered. This "pyramid of biomass", in effect, gives a picture of the over-all effect of food chain relations for an ecological group as a whole.

Pyramid of Energy A graphical representation of the rate of energy flow and/or energy productivity at successive trophic levels in a food chain or for a complete biotic community. Of the three types of ecological pyramids, the energy pyramid gives the best overall picture of the functional nature of biotic communities since the number and weight of organisms that can be supported at any level in any situation depend not on the amount of fixed energy present at any one time in the level just below, but rather on the rate at which food is being produced. In contrast the pyramid of numbers and pyramid of biomass, are pictures of the standing states, i.e., organisms present at any one moment, while the energy pyramid is a picture of the rates of passage of food mass through the food chain. Its shape is not affected by variations in the size and metabolic rate of individuals, and, if all sources of energy are considered, the largest numbers will always be at the base because of the second law of thermodynamics.

Pyramid of Energy

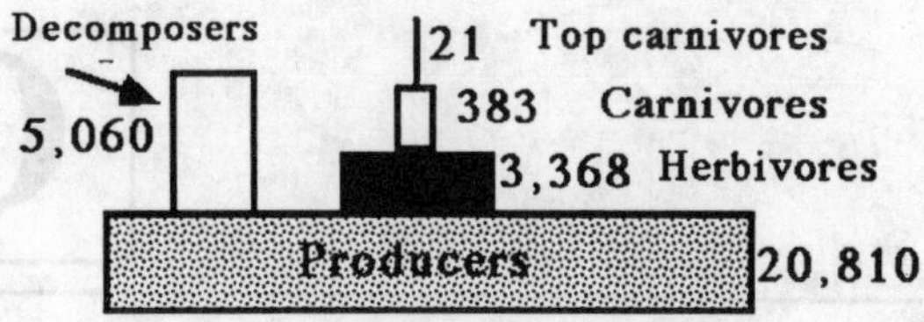

Kilocalories /meter square/year

Fig. Pyramid of Energy

Pyramid of Numbers (Eltonian Pyramid, Food Pyramid) A graphic representation of the food chain which indicates the large numbers of producer organisms (i.e., plants) at the base of the food chain, and the progressively decreasing numbers of herbivores (plant eaters) and carnivores (meat eaters) higher up in the food chain. Elton's concept was that in most food chains the number of individuals decreases in each succeeding stage, with large numbers of animals occurring at the base and a few large ones at the top.

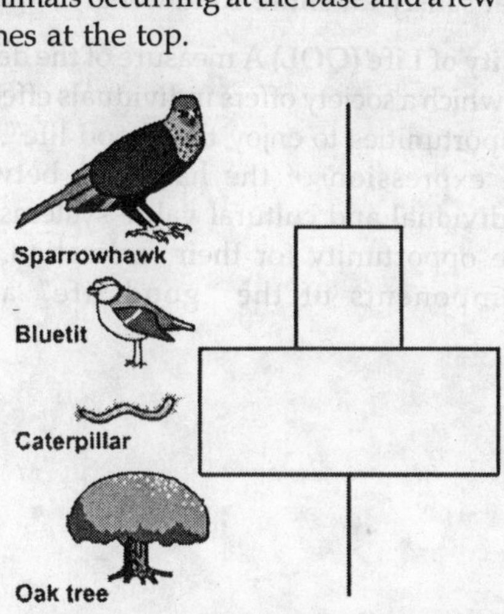

Fig. Pyramid of Numbers

Q

Q Factor A term used in singletree selection to describe the distribution of size classes. This is expressed in terms of the ratio q between numbers of trees in successive 2-inch diameter classes. For example, a q of 1.5 means that there are 1.5 times as many 10-inch trees as there are 12-inch trees; 1.5 times as many 12-inch trees as there are 14-inch trees. The lower the q ratio, the more large trees there are in proportion to small trees.

Qualified Data Records Those data records that satisfy an entire conditional clause.

Quality of Life (QOL) A measure of the degree to which a society offers individuals effective opportunities to enjoy the "good life". It is an expression of the harmony between individual and cultural value systems and the opportunity for their realization. The components of the "good life" are a combination of goods and services available in a total environment, one that is physical, social, and cultural. Their precise composition is a matter of personal judgement about which widely differing opinions are held. The "quality of life", therefore, is a subjective concept. It may, however, be that, for a particular community, a reasonable consensus could be found about a range of choice of goods and services and about the qualities of the environment that should be offered. Giles argued that when a well-specified set of objectives is approximately achieved, then a high quality of life exists. Thus, QOL is an expression of the degree of achievement of objectives.

Questionnaire A set of questions for obtaining useful information about or from a person or group.

R

R&D Research and Development.

Radiant 1. In optics, the point or object from which light proceeds.

2. In geometry a straight line proceeding from a given point, or fixed pole, about which it is conceived to revolve.

3. In astronomy the point in the heavens from which a shower of meteors seems to proceed.

Radiation (Solar) The process (or the radiated energy itself) in which energy is emitted from the sun, transmitted through space, and received by Earth.

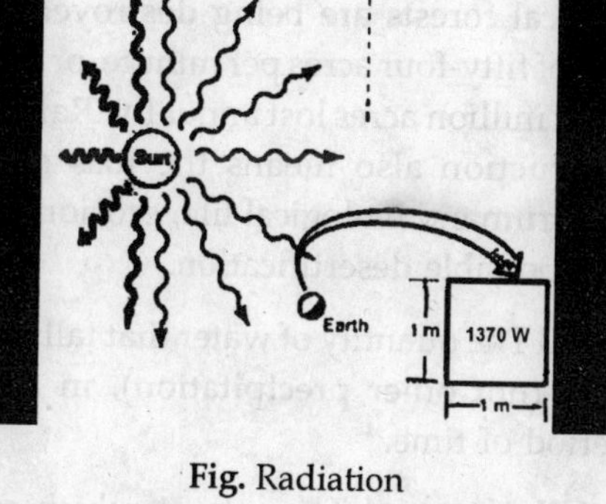

Fig. Radiation

Radiation Energy transfer in the form of electromagnetic waves or particles that release energy when absorbed by an object.

Radiation budget A measure of all the inputs and outputs of radiative energy relative to a system, such as Earth.

Radiative cooling Cooling process of the Earth's surface and adjacent air; which occurs when infrared (heat) energy radiates from the surface of the Earth upward through the atmosphere into space. Air near the surface transfers its thermal energy to the nearby ground through conduction, so that radiative-cooling lowers the temperature of both the surface and the lowest part of the atmosphere.

Radiative transfer Theory dealing with the propagation of electromagnetic radiation through a medium.

Radio frequency (RF) A frequency that is useful for radio transmission, usually between 10 kHz and 300,000 MHz.

Radio spectrum The complete range of frequencies or wave lengths of electromagnetic waves, specifically those used in radio and television.

Radio wave An electrical impulse sent through the atmosphere at radio frequency.

Radioactive Giving off or capable of giving off radiant energy in the form of particles or rays, as in alpha, beta, and gamma rays.

Radiochemical Species Refers to the individual radioactive elements that produce radioactivity, such as radium 226, cobalt 60, strontium 90, and tritium.

Radiometer An instrument that quantitatively measures electromagnetic radiation. Weather satellites carry radiometers to measure radiation from snow, ice, clouds, bodies of water; the Earth's surface, and the sun.

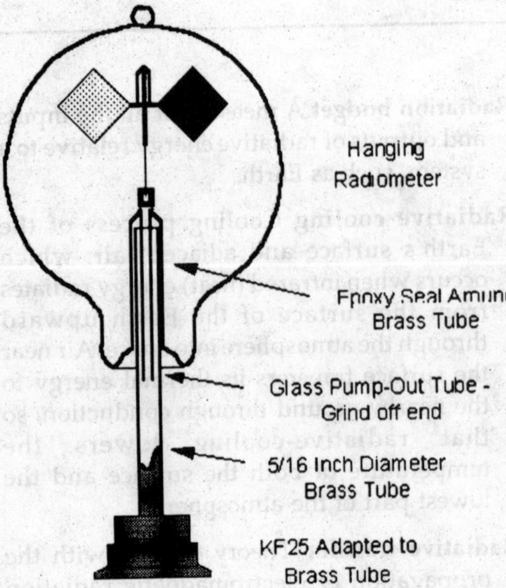

Fig. Radiometer

Radiosonde A balloon-borne instrument that measures meteorological parameters from the Earth's surface up to 20 miles in the atmosphere. The radiosonde measures temperature, pressure, and humidity and transmits or "radios" these data back to Earth. Upper air winds also are determined through tracking of the balloon ascent.

Radiosonde observations generally are taken twice a day (0000 and 1200 UTC) around the globe. NOAA's National Weather Service (NWS) operates a network of about 90 radiosonde observing sites in the U.S and its territories. When the balloons burst radiosondes return to Earth on a parachute. Approximately 25 percent are recovered and returned to NWS for reconditioning and reuse.

Rain forest An evergreen woodland of the tropics distinguished by a continuous leaf canopy and an average rainfall of about 100 inches per year. Rain forests play an important role in the global environment. The Earth sustains life because of critical balances and interactions among many factors. Were there not processes at work that limit the effects of other essential processes, Earth would become uninhabitable. Destruction of tropical rain forests reduces the amount of leaf area in the tropics, and consequently the amount of carbon dioxide absorbed, causing increases in levels of carbon dioxide and other atmospheric gases. It is estimated that cutting and burning of tropical forests contributes about 20 percent of the carbon dioxide added to the atmosphere each year. The World Resources Institute and the International Institute for Environment and Development have reported that the world's tropical forests are being destroyed at the rate of fifty-four acres per minute, or twenty-eight million acres lost annually. Rain forest destruction also means the loss of wide spectrumand biological life, erosion of soil, and possible desertification.

Rainfall The quantity of water that falls as rain only (not other precipitation), in a given period of time.

Rainfall Intensity The rate at which rain is falling at any given instant, usually expressed in inches per hour.

Rain gauge Calibrated container that measures the amount of rainfall during a specific period of time.

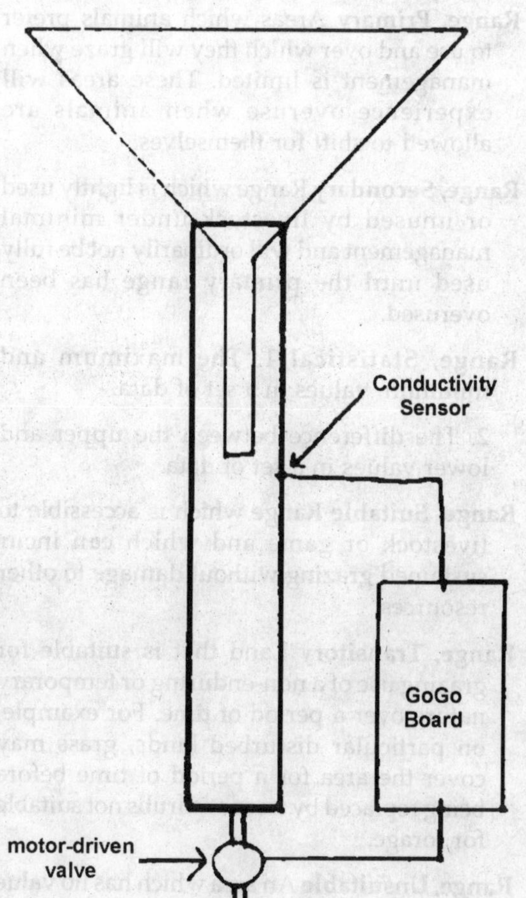

Fig. Rain gauge

RAM Random Access Memory Computers use two types of memory RAM and ROM. RAM is the computer's working area, the primary location where the microprocessor stores the information it needs. The designation "random access" stems from the microprocessor's ability to access information in memory randomly by knowing its location, or address, rather than hunting through memory sequentially from beginning to end. Because information in RAM is stored electronically accessing data stored in RAM is much faster than getting that data from a mechanical storage device such as a disk drive. But because it is stored electronically all information in RAM is temporary (which is why you must store it on a more permanent storage capability, such as a disk). Compare with ROM.

Range All land producing naturalized or native forage for animal consumption, and lands that are revegetated naturally or artificially to provide a forage cover that is managed like naturalized or native vegetation. Generally considered as land that is not cultivated. It may include forestlands that support an understory or periodic cover of herbaceous or shrubby plants suitable for grazing without impairing other forest values.

Land on which the native vegetation is predominately grasses, grass-like plants, forbs, or shrubs suitable for grazing or browsing use. Forested sites and nonforested sites providing forage and habitat for domestic and wild herbivores are included (US Forest Service, 2005).

Range Allotment (Allotment) An area designated for the use by a prescribed number of cattle or sheep, or by common use of both under one plan of management.

Range Condition and Trend 1. Range condition — the status of vegetation and soil of a given range area in relation to the optimum status (considered by some to be climax community) obtainable under the prevailing environmental conditions.

2. The terms range *condition and trend* are used to express the idea that at any given moment in time a vegetation unit assumes a specific set of characteristics (condition) and that these characteristics change (trend).

Range Condition Class One of a series of arbitrary categories used to classify range condition, usually expressed as either excellent, good, fair, or poor.

Range Condition The state and health of the range based on what it is naturally capable of producing.

Range Improvement Any structure or excavation to facilitate managing of range or livestock (physical) or an increase in the grazing capacity or condition of range (biological).

Range Management Making decisions and taking action to direct and manipulate range use to obtain sustained maximum profitable animal production and other stated benefits, consistent with perpetuating the natural resources of rangelands.

Range of Variation Spatial and temporal variation in ecosystem characteristics during a period of time when the influences of European-American settlement were minimal.

Range Readiness The stage of growth of the important palatable plants on the range and the condition of soil which permit grazing without animals unduly compacting the soil or endangering the ability of the plants to maintain themselves.

Range Site An area of land having a combination of soil, climatic, topographic and natural biotic factors that result in plants significantly different from those on other nearby areas.

Range Suitability The adaptability of a range to grazing by livestock and/or game.

Range, Annual Range on which the principal forage plants are herbaceous species which only live for one year.

Range, Closed A suitable range that is closed to grazing by domestic livestock.

Range, Key Winter The small portion of the total year's range where big game animals find food and/or cover during severe winter weather.

Range, Open All suitable range of an area upon which grazing is permitted, and typically those which have not been fenced into management units.

Range, Primary Areas which animals prefer to use and over which they will graze when management is limited. These areas will experience overuse when animals are allowed to shift for themselves.

Range, Secondary Range which is lightly used or unused by livestock under minimal management and will ordinarily not be fully used until the primary range has been overused.

Range, Statistical 1. The maximum and minimum values in a set of data.

2. The difference between the upper and lower values in a set of data.

Range, Suitable Range which is accessible to livestock or game and which can incur sustained grazing without damage to other resources.

Range, Transitory Land that is suitable for grazing use of a non-enduring or temporary nature over a period of time. For example, on particular disturbed lands, grass may cover the area for a period of time before being replaced by trees or shrubs not suitable for forage.

Range, Unsuitable An area which has no value for, or which should not be used by, livestock because of steep topography, barrenness, dense timber, lack of forage, or unstable soils.

Rangeland Grazing System All of those activities that bear directly upon managing, protecting, and developing the land for livestock or wildlife grazing. It also includes the cooperative aspects of state and private forestry associated with producing livestock. Research provides the system with management alternatives that relate range productivity with other forest and range land resources.

Rangeland Land, a major use of which has been grazing by livestock and big game animals. It is land on which the potential natural vegetation (climax) of plants is dominated by native grasses, grass-like plants, forbs, and shrubs. Some rangelands have been or may be seeded to introduced or domesticated plant species but are managed like native vegetation. Rangelands include natural grasslands, savannas, shrublands, many deserts, tundra, alpine communities, coastal marshes, and wet meadows. It includes facilities and areas of water but is generally land with less than 25 percent tree canopy or less than 10 percent stocking with forest trees of any size.

Rangeland Vegetation Vegetation on all land with rangeland resource objectives or rangeland resource values, including riparian areas. Generally, the focus is on land supporting grass or grass-like plants, forbs, or shrubs during one or more ecological stages. Forested and non-forested sites providing forage and habitat for wild and domestic animal species are included.

Ranging Engaging in one or more of a diverse set of extended, dispersed outdoor or rural activities for health, recreation, study, appreciation, and adventure. It may also mean the enterprises that are related and that promote, support, and supply these activities and the areas and resources used. It includes (but is not limited to) hiking, tramping, trecking, camping, rock and tree climbing, biking, horseback trail riding, hunting, fishing, canoeing, kyacking, boating, touring, sightseeing, studying nature, and active wildlife watching. A word like ranging can help provide structure to the confusing array of inconsistent writing and work in the expansive areas such as *outdoor recreation, adventure tourism, nature tourism,* and *ecotourism*. Ecotourism is for visitors to an area; ranging is for residents as well as visitors.

Raw Land (Undeveloped Land) Undeveloped land or land not subdivided in contrast to land which has been built upon (i.e., "developed") or subdivided in preparation for development.

Real Dollar Value A monetary value that compensates for the effects of inflation.

Receipt Shares The portion of receipts derived from central land resource management that is distributed to State and county governments.

Receiver sensitivity The ability of a receiver to detect weak signals through the noise level of the receiving system, which includes the antenna and internal thermal noise of the receiver.

Recharge The addition of water (or the water itself) to an aquifer that occurs naturally from infiltration of rainfall and from water flowing over Earth materials.

Reclamation 1. The process of converting disturbed lands or mined land to its former (or other) productive uses or form, e.g., filling in of open pits, grading of the mined area, reducing high walls, replacing topsoil, planting, and revegetating.

2. Recovery to a state of closure of nutrient pathway loops that are as closed (or open) as before disturbance through all of the range of geologic and climatic conditions to which the site is potentially subject.

3. The process of making a site habitable to organisms that were originally present or others like the original inhabitants.

4. Returning disturbed lands to a form and productivity that will be ecologically balanced and in conformity with the predetermined land-management plan.

Reclamation, Effective Rehabilitating mining-affected land to make it acceptable for designated purposes, with protection of the natural resources and aesthetic value of adjoining areas.

Reconnaissance Survey A preliminary visit and report usually executed rapidly and at a relatively low cost.

Record of Decision A document separate from, but associated with, an environmental impact statement that publicly and officially discloses the responsible official's decision as to which alternative assessed in the EIS is to be implemented.

Recreation Use of leisure time for personal satisfaction and enjoyment and for physical and mental health. Recreation may be a basic human need. It may be undertaken individually or with others. It may be planned or spontaneous. It may be passive or active, may or may not require skills and training, and may or may not require a designated area. It includes such activities as swimming, picnicking, boating, hunting, and fishing. It is not a resource but a complex activity composed of people (the recreationists), the environments, and their actions. Recreation, for some people, exists in the mind and takes place in an environment. Outdoor recreation involves protecting, preserving, developing, using, and enjoying scenery, water, primitive or natural landscape (including roadless areas), wildlife, natural phenomena (e.g., petrified wood), and archeological and historical sites.

Recreation Area, Developed A relatively small, distinctly defined area where concentrated public use for the more traditional recreation purposes predominates, e.g., campgrounds, picnic areas, swimming areas.

Recreation Composite An identified geographical area containing special, significant and needed outdoor recreation resources, features, or opportunities and concentrations of land. They may be planned as a composite.

Recreation Day A standard unit of use for an area or activity consisting of a visit by one individual to a recreation development or area for recreation purposes during any reasonable portion or all of a 24-hour period. This corresponds roughly with what is termed a "recreation visit". For example, if the average stay at a picnic site were four hours, then a four-hour visit to the picnic site by one person would be a recreation day.

Recreation Land Land and water used primarily for recreation. (Recreation may be a secondary use of land where woodland, cropland, rangeland, etc., is the primary use).

Recreation Opportunity Spectrum (ROS) Land delineation that identifies a variety of recreation experience opportunities categorized into six classes on a continuum from primitive to urban. Each class is defined in terms of the degree to which it satisfies certain recreation experience needs, based on the extent to which the natural environment has been modified, the type of facilities provided, the degree of outdoor skills needed to enjoy the area, and the relative density of recreation use. The six classes are:

1. Primitive - Area is characterized by an essentially unmodified natural environment of fairly large size. Interaction between users is very low and evidence of other users is minimal. The area is managed to be essentially free from evidence of human-induced restrictions and controls. Motorized use within the area is not permitted.

2. Semiprimitive Non-motorized - Area is characterized by a predominantly natural or natural-appearing environment of moderate to large size. Interaction between users is low, but there is often evidence of other users. The area is managed in such a way that minimum on-site controls and restrictions may be present but would be subtle. Motorized recreation use is not permitted, but local roads used for other resource management activities may be present on a limited basis. Use of such roads is restricted to minimize impact on recreational experience opportunities.

3. Semi-primitive Motorized - Area is characterized by a predominantly natural or natural appearing environment of moderate to large size. Concentration of users is low, but there is often evidence of other users. The area is managed in such a way that minimum onsite controls and restrictions may be present but would be subtle. Motorized recreation use of local primitive or collector roads with predominantly natural surfaces and trails suitable for motor bikes is permitted.

4. Roaded Natural - Area is characterized by predominantly natural appearing environments with moderate evidence of the sights and sounds of man. Such evidence usually harmonizes with the natural environment. Interaction between users may be moderate to high, with evidence of other users prevalent. Resource modification and utilization practices are evident but harmonize with the natural environment. Conventional motorized use is allowed and incorporated into construction standards and design of facilities.

5. Rural - Area is characterized by a natural environment that has been substantially modified by development of structures, vegetative manipulation, or pastoral agricultural development. Resource modification and utilization practices may be used to enhance specific recreation activities and to maintain vegetative cover and soil. Sights and sounds of humans are readily evident, and the interaction between users is often moderate to high. A considerable number of facilities are designed for use by a large number of people. Facilities are often provided for special activities. Moderate user densities are present away from developed sites. Facilities for intensified motorized use and parking are available.

6. Urban - Area is characterized by a substantially urbanized environment, although the background may have natural-appearing elements. Renewable resource modification and utilization practices are often used to enhance specific recreation activities. Vegetative cover is often exotic and manicured. Sights and sounds of humans are predominant onsite. Large numbers of users can be expected both onsite and in nearby areas. Facilities for highly intensified motor use and parking are available with forms of mass transit often available to carry people throughout the site.

Recreation Resource The combination of the natural qualities of land and water areas and the ability and desire of people to use this combination for recreation.

Recreation, Concentrated Those types of activities associated with developed recreation sites, e.g., such as campgrounds, picnic grounds, ski areas, fishing ramps, scenic overlooks, interpretive sites.

Recreation, Dispersed (Extensive) Recreation of various kinds that occurs, largely individual, scattered, generally throughout a large area, and is not confined to a specific place or developed facilities. Examples are recreational use of trails, natural lakes, ponds, rivers, streams and generally undeveloped areas. It includes such activities as hiking, backpacking, hunting, fishing, using snowmobiles, horseback riding, and cross-country skiing.

Recreation, Intensive High-density recreational activities, e.g., developed camp and picnic grounds, swimming beaches, hiking trails that are major access routes into wilderness areas, all-terrain vehicle areas.

Recreation, Primitive Those types of recreational activities associated with unroaded land, e.g., hiking, backpacking, and cross-country travel.

Recreational Experience Levels The extent to which various classes of outdoor recreation experiences provide opportunities for satisfying some of the basic needs of individuals, e.g., to find isolation and solitude, to socialize, to achieve self-fulfillment, for aggression outlets. There are 6 classes based upon the extent to which; 1. Outdoor skills are required,.

2. The natural environment has been modified to facilitate use and,.

3. Comfort and/or convenience facilities are provided.

Recreational Experience Quality The degree of excellence, value, or worth of any form of recreational experience determined through personal judgments and preferences. Consensus and individual variation form expectations that, depending upon the extent to which they are achieved, specify the degrees of quality judgments. Consensus may provide relative value scales or standards that may offer some universal basis for quality judgments. Individual variation, as shown through personal judgments and preferences, is the deviation from consensus as a consequence of individual intelligence and qualities of personality as influenced by education, experience, perception, and socialization and modified by external influences such as area, design facilities, activities, human relations, impacts, finances, health and safety, weather, travel, and time.

Recreational Quality The degree to which a recreational experience or area normally contributes to the physical and psychic well being of recreationists. Thus, a thoroughly worn area usually offers lower recreational quality than areas presenting a fresh and natural scene. Individual tastes differ, and an experience is of high quality only to people for whom it provides a large measure of enjoyment or well being.

Recreational River Area Those rivers or sections of rivers that are readily accessible by road or railroad, that may have some development along their shorelines, and that may have undergone some impoundment or diversion in the past.

Recursive The process of repeating an operation or group of operations, usually with the result of each repetition being in some way dependent upon the result of the preceding repetition, e.g., planning itself and ecological succession within a rotation.

Reference Landscapes Terrestrial and aquatic areas with high ecosystem integrity and within the historical range of variability and of sufficient size, where relevant disturbance and ecological processes occur and are generally unaffected by human activities.

Reflection The return of light or sound waves from a surface. If a reflecting surface is plane, the angle of reflection of a light ray is the same as the angle of incidence.

Reforestation Natural or artificial restocking or reestablishing an area with forest trees (natural regeneration as well as tree planting and seeding); renewing forest cover by planting seedlings, transplants, tree seeds, or, (for certain species) cuttings.

Reforestation Backlog Areas that need to have trees reestablished. Planting, seeding, or preparing the site for natural regeneration can do this.

Refuse All of the solid waste from a coal mine, including tailings and slurry. Other synonyms are dirt, gob, shale, and slate.

Regeneration (Reproduction) 1. Renewing a forest after harvest whether by natural or artificial means including natural reproduction by self-sown seed, by vegetative means such as coppicing, or by planting or direct seeding.

2. The young tree crop itself, i.e., seedlings and saplings.

Regeneration Cut Removing trees with the intention of establishing a new forest stand.

It may be a clearcut, seed tree cut, shelterwood cut, or selection cut.

Regeneration Cutting 1. Any removal of trees intended to assist regeneration already present or to make regeneration possible. "Regeneration cutting" applies generally to the logging stands of rotation age or greater (virgin stands and to all cutting in cutover stands where the overwood components of the stand are over rotation age), and stands below rotation age which cannot economically be held any longer because of poor stocking, health, thrift, quality, or composition.

2. The final cut made in a stand with the intent to create a new stand.

Region A defined portion of the Earth's surface distinguished by certain natural features, climatic conditions, a special fauna or flora of a definite extent or character. An area with unity and different from adjacent regions.

Region, Morphogenetic (Formkreis) A region with distinguishing landscape characteristics which are the result of particular dominant geomorphic processes themselves with origins in a set of climatic conditions.

Region, Morphological Land units with a high degree of homogeneity in all their morphological (i.e., land surface form) features. Areas possessing a uniformity of land surface forms without any implied, underlying speculation as to the processes responsible for those forms.

Region, Natural A part of the Earth's surface with a comparatively high degree of uniformity of structural and climatic features within it. "Natural regions" show mutual adjustment of drainage and landforms, but also well-marked zones of climate, vegetation, and even human distributions that characterize such forms when situated in similar climatic areas. A natural region should have a certain unity of configuration, climate, and vegetation. The basis of the regional division is to be found in nature as a whole, including people, in contrast to a division based on a single element, as in the case of "climatic regions," "agricultural regions" etc. The term "natural region" has been used to cover two distinct types of unit-areas of the earth's surface: 1. Those which are marked out as possessing certain common physical characteristics – e.g., a certain kind of structure and surface relief, or a particular kind of climate; and.

2. Those regions which possess a unity based upon any significant geographical characteristics whether physical, biological or human, or any combination of these, as contrasted with areas marked out by boundaries imposed, frequently for political or administrative purposes, without reference to any geographical unity of the areas.

Region, Physiographic A region having a particular pattern of relief features and climate or land forms that differs significantly from that of adjacent regions.

Regional Climate (Macroclimate) The general large-scale climate of a large area or country, as distinguished from the meso-climate and micro-climate.

Regional Economic Development One of the four "required accounts" for categorizing, displaying, or "accounting" the beneficial and adverse effects of each alternative plan formulation for water and related land resources planning.

Regional Plan The plan for a region (i.e., an area larger than a single city or county unit and whose boundaries have typically been designated according to some physiographic, biological, political, administrative, economic, demographic or other criteria).

Regional Planning The planning discipline concerned with any area larger than a single city. Regional planning is a discipline, primarily oriented to the future, which seeks to relate social purposes and resource use and development in any large area.

Regolith The layer or mantle of loose, non-cohesive, or cohesive rock material, of whatever origin, that nearly everywhere forms the surface of the land and rests on bedrock. It is comprised of rock waste of all sorts; volcanic ash; glacial drift, alluvium; wind-blown deposits; and accumulations of vegetation, such as peat; and soil.

Regression A statistical method for studying and expressing the change in one variable associated with and dependent upon changes in another related variable or group of variables. A line that shows the relationships between two variables.

Regulated Acres Commercial forest land organized for sustained yield production of wood products.

Regulated Forest A forest that has reached a level where the cut equals the growth. Barring unforeseen disasters, a regulated forest will produce a continuous even flow of products forever, if each stand is harvested when it becomes mature and is replaced by a new stand immediately thereafter.

Regulated Forest land managed for timber production under sustained yield principles.

Regulated Harvest The volume scheduled in calculations of the allowable sale quantity which is harvested from suitable commercial forest land.

Regulation 1. The artificial manipulation of streamflow;.

2. A rule established to make laws more precise and to guide actions.

Rehabilitation 1. Implies that the land will be returned to a form and productivity in conformity with a prior land use plan, including a stable ecological state that does not contribute substantially to environmental deterioration and is consistent with surrounding aesthetic values.

2. The process of returning a disturbed site to its original ecological slate.

Rehabilitation Potential The probability of returning land that has been surface-mined to a use which is compatible with the existing environment.

Release Cutting Freeing a tree or group of trees, from competition by cutting, or otherwise eliminating growth that is overtopping or closely surrounding them. Three operations come: 1. Treatments to free desired growing stock trees from over-story vegetation; principally, scrub trees, brush, and vines that inhibit their growth or desirable development.

2. Treatments to prevent development of brush and young trees that threaten to overtop desired growing stock trees; or.

3. Removal of grass, weeds, or brush from around individual seedlings or small trees to release them from competition for soil moisture, shade, or smothering as under matted grass or weeds.

Relic (Relict) Something that is left behind after decay, disintegration, or disappearance (e.g., a structure, a culture, etc.). A community, fragments of presently existing flora on areas with a historical record of no disturbance and the community is assumed to be in near virgin condition. A community or species which, through the operation of some compensating or protective environmental features has survived some major change (e.g., climatic, land use) that has altered the general vegetation and ecology of the surrounding territory.

Relief The difference in elevation between the high and low points of a land surface or the inequalities of a land surface, taken collectively.

Remapping Flattening the Earth into a standard map projection. When the spherical Earth is photographed by satellites, areas lying near the outer edge of the picture are distorted. Remapping rectifies the distortion.

Remote sensing The technology of acquiring data and information about an object or phenomena by a device that is not in physical contact with it. In other words, remote sensing refers to gathering information about the Earth and its environment from a distance, a critical capability of the *Earth Observing System.*

For example, spacecraft in low-Earth orbit pass through the outer thermosphere, enabling direct sampling of chemical species there. These samples have been used extensively to develop an understanding of thermospheric properties. Explorer-17, launched in 1963, was the first satellite to return quantitative measurements of gaseous stratification in the thermosphere. However, the mesosphere and lower layers cannot be probed directly in this way—global observations from space require remote sensing from a spacecraft at an altitude well above the mospause. The formidable techological challenges of atmospheric remote sensing, many of which are now being overcome, have delayed detailed study of the stratosphere and mososphere by comparison with thermospheric research advances.

Some remote-sensing systems encountered in everday life include the human eye and brain, and photographic and video cameras.

Removal Cut Cutting of the trees left as a source of seed for natural regeneration after the new stand is established.

Removal Regeneration Cutting Cutting a stand which has an advanced manageable understory of 1-20 year old age class reproduction. The seed-tree cutting is bypassed and all the overstory competition is removed in one cutting. The stand is reduced to the single age class for future management.

Renewable Resource Lands Areas where uncontrolled or incompatible development which results in the loss or reduction of continued long range productivity could endanger future water, food, and fiber requirements of more than local concern. These lands include watershed lands, aquifers and aquifer recharge areas, significant agricultural and grazing lands, and forest lands.

Renewable Resource Resources, the supply of which becomes available for use at different intervals in time. The use of present supply flows does not diminish future flows and it is possible to maintain use indefinitely provided the use rate does not exceed flow rate. Many renewable resources can be stored — such as water in a reservoir or mature timber in a stand and such stores can be treated as a stock (i.e., exhaustible) resource. The availability of a flow for replenishing the stock, however, differentiates such stores from truly "nonrenewable resources." There are two broad classes of "renewable resources" one in which human resource use activities do not significantly alter future flows (e.g., wind, tidal and solar energy) and a second class where human activities may increase or decrease future flows (e.g., timber, wildlife, and soil productivity).

Renewable Resources Assessment An appraisal of the Nation's renewable resources that recognizes their vital importance and the necessity for long-term planning and associated program development. The Assessment meets the requirements of Section 3 of the Forest and Rangelands Renewable Resources Planning Act (RPA) of 1974 and includes analyses of present and anticipated uses, demands, and supplies of the renewable resources; a description of Forest Service programs and responsibilities; and a discussion of policy considerations, laws, and regulations.

Renewable Resources Program The program for managing and administering the National Forest System, for Research and Development, for Cooperative State and Private Forest Service programs, and for conducting other Forest Service activities developed in accordance with the Forest and Rangeland Renewable Resources Planning Act of 1974.

Repository A place of storage or safekeeping; an archive, e.g., water data in a data bank.

Research Natural Area Land areas containing natural plant communities that have been little modified by people, and are preserved, protected, and studied as a representative sample of an ecological community primarily for scientific and educational purposes. These are physical and biological units in as near natural condition as possible, exemplifying typical or unique vegetation and associated biotic, soil, geologic and aquatic features. Commercial use is not allowed and public use is discouraged.

Reservoir A pond, lake, or basin, either natural or artificial, for the storage, regulation, and control of water.

Residential Density The average number of families living on one acre of land in a given area. "Gross residential density" is obtained by dividing all land in a defined area used for residences, streets, local schools, local parks and local shopping facilities into the total number of families in an area. "Net residential density" is determined by dividing the total number of families in a defined area by the total acreage of all parcels of land within the area that is used for residential purposes.

Residential Land Land used primarily for permanent dwellings, such as houses, apartments, and housing developments, including adjacent facilities such as underground and surface utilities, access streets and alleys, servicing structures, appurtenances, and measures.

Residual Value Appraisal A timber appraisal that subtracts logging, transportation, and manufacturing costs from the value of finished products to get an estimated stumpage price.

Resolution A measure of the ability to separate observable quantities. In the case of imagery, it describes the area represented by each pixel of an image. The smaller the area represented by a pixel, the more accurate and detailed the image. APT has a resolution of 4 km, i.e., each pixel represents a square, 4 km on each side. HRPT has a resolution of 1.1 km at nadir (4 km at edge of scan), and WEFAX of 8 km.

Resolution cell The smallest unit of area in an image of discrete elements. The area represented by a pixel.

Resource 1. A human-valued commodity or product provided through management of the land. Resources may or may not have any commercial value. They include timber, wildlife, water, minerals, recreation, scenery, aesthetics, and cultural.

2. A source of supply.

3. Anything useful to people directly or that facilitate use and satisfaction of human wants, be it animal, vegetable, mineral, a location, a labour force, etc. Resources vary from commodities such as timber and minerals to amenities as scenery or scenic viewing points. "Resource" is a highly relative concept, the precise definition of which changes depending upon the planning agent and the planning purpose. For example, garbage is not a "resource" to those attempting to get rid of their wastes

but it is a "resource" to those who are attempting to recover recyclable materials from it, provided existing regulations and separation technology permit them to do this economically.

Resource Allocation Resource Allocation - The decisions and actions to apportion and distribute available resources to various activities, people, or programs in such a way that total effectiveness is likely to be increased or maintained. Difficult allocation decisions are necessary when there are limitations (constraints) on either the amount of resources available or on the way on which these may be expended, especially when each separate activity cannot have resources sufficient for maximally effective performance.

Resource Management Prescriptions Written direction for the resource management practices selected and scheduled for application on a specific area to attain goals and objectives.

Resource Management System (RMS) A combination of interrelated conservation, practices and management techniques used to maintain or improve soil, water, plant and related resources. The system is designated according to primary use.

Resource Plan A plan prepared by a unit(s) of government and used as a guide for public policy and action on the use, treatment, and management of soil and water resources. Resource plans are generally parts of comprehensive, regional, or functional plans.

Restoration The process of restoring site conditions as they were before a land disturbance.

Retention 1. A visual quality objective which in general means human activities are not evident to the casual forest visitor.

2. The amount of precipitation on a drainage area that does not escape as runoff. It is the difference between total precipitation and total runoff.

Revegetation Reestablishing and developing a plant cover. This may take place naturally through the reproductive processes of the existing flora or artificially through the direct action of people, e.g., afforestation, range reseeding.

Revenue-Cost Analysis A method of comparing alternatives by analyzing the monetary income that each alternative would generate in relation to its cost.

Revolution Process of the Earth circling the sun in its orbit. Revolution determines the seasons, and the length of the year. In addition, differences in seasons occur because of Earth's inclination (tilt on its axis) of about 23.5 degrees as it revolves around the sun. Compare with rotation.

Right of Eminent Domain The taking of property for a necessary public use, with reasonable compensation being made to the property owner.

Right-justified A string of characters, such as a title, positioned within a column in such a way that its last character falls on the right most part of the column.

Right-of-way (ROW) Strictly, it is the right of passage over another person's ground. In its legal and generally accepted meaning, in reference to a roadway, it is a mere easement in the lands of others, obtained by lawful condemnation for public use or by purchase. It is unusual to use the term to apply to an absolute purchase of the fee simple ownership of land to be used for a roadway or other kind of way. "Right-of-way" has two-fold significance being sometimes used to mean the mere intangible right to cross, and often used to otherwise indicate that strip of land upon which a road, powerline, etc. is built.

Riparian Land adjacent to perennial streams, lakes, and reservoirs and including other well developed riparian vegetation (primarily intermediate streams). This land is specifically delineated by the transition between the aquatic ecosystem and the adjacent terrestrial ecosystem and defined by soil characteristics and distinctive vegetation communities that require free and unbound water. Areas include stream channels, lakes, wetlands, and adjacent floodplains and riparian ecosystems. They include all areas within a horizontal distance of 100 feet from the edge of perennial streams or other water bodies.

Riparian Dependent Resource Any resource that is dependent on a riparian area (even for a short time) for its existence.

Riparian Ecosystems The transition between the aquatic ecosystem and the adjacent terrestrial ecosystem. It is identified by the presence of very poorly drained soils (alluvial soils inundated by select floods and wetland soils).

Riparian Rights The rights of owners of lands on the banks of watercourses or bodies, relating to the water, its use, ownership of soil under the stream, shoreline accretions, etc. These rights vary with state law. Riparian rights cease at the water's edge and do not interfere with use of the water area by others offshore.

Riparian Type A systematic method of characterizing streams by evaluating form, bottom type, gradient, sidè slopes, flood plain, and channel condition.

Riprap Material such as rock, logs, and concrete which is placed along a watercourse to stabilize the banks.

Risk Cutting Cutting to remove trees that are likely to die before the next periodic cut.

Risk Uncertainty in a situation is characterized as risk if it is possible to describe all possible outcomes and to assign meaningful, objective, numerical, probability weights to each one. For example, an action might lead to this risky outcome: a reward of $10 if a "fair" coin comes up heads, and a loss of $5 if it comes up tails. Under the Bayesian school of thought the distinction between "risk" and uncertainty is not useful and in fact may not exist. In any practical analysis all uncertain situations are as signed probability distributions, either explicitly or implicitly by the selection of the analysis method. Whether or not the true probability distribution is known, a probability distribution is selected for analysis purposes based on experience, intuitive feelings, knowledge, and the lack of knowledge. Typically since little can be proven, risk treats failure, i.e., the possibility of loss, injury, disadvantages, or destruction.

Other observers combine concepts, defining it as a combination of the likelihood that a negative outcome will occur and the severity or magnitude of the subsequent negative consequences.

Risk, Acceptable The level of risk below which no specific action by local government is deemed necessary, other than making the risk known.

Risk, Avoidable A risk not necessary to be taken because the individual or public goals can be achieved at the same or less total "cost" by other means without taking the risk.

Risk, Unacceptable The level of risk above which specific action by a manager or government is deemed necessary to protect life, property, and achievement of objectives.

River Basin Plan (Regional Plan) One of the "levels of planning" for water and related land resources. A plan for development of water and related land resources to make the best use of such resources to meet the basin needs and make the greatest long-term contribution to the economic growth and social well being of the people of the basin and the nation. "Regional" or "river basin plans" are reconnaissance-level evaluations of water and land resources for selected areas. They are directed toward resolving the complex problems identified by framework studies and assessments and terminate in a recommended plan or disclosure of possible alternative plans. They may vary widely in scope and detail; will consider present and long-range problems with a focus on middle term (15 to 25 years) needs and desires; and will involve interested central, state, and local entities. Regional or river basin plans are concerned with a broad array of component needs of objectives. Alternate plans will consider effects on many components of objectives. Scheduling for implementation of the recommended plan or alternative plans will be presented to indicate the relationship to needs and the urgency or priority associated with meeting the needs.

Riverine Includes all wetlands and deepwater habitats contained within a channel. A channel is an open conduit either naturally or artificially created which periodically or continuously contains moving water. It is bounded on the landward side by upland, by the channel bank, or by wetland dominated by trees, shrubs, persistent and emergents. It terminates at the upstream end where tributary streams originate and at the downstream end where the channel enters a lake.

Road Closed Any road that has an operative closure device, such as a gate, a ditch, or earthen mound.

Road Density The measure of the degree to which the length of road miles occupies a given land area, i.e., miles/acre or km/ha.

Road For the purposes of wilderness inventory, "roadless" means absence of roads which have been improved and maintained by mechanical means to insure relatively regular and continuous use. A way maintained solely by the passage of vehicles does not constitute a road.

Road Influence The effect a road has on animal distribution, behaviour, and vulnerability to hunters. A zone of influence often associated with calculations involving habitat effectiveness.

Road Open A road that is open to the public. All types of traffic are permitted, but may be restricted when necessary to protect the road from damage or for public safety.

Road Restricted Any road that has restrictions on the type of use such as size limitations on weight, or vehicle types including restrictions on public use.

Roadless and Undeveloped Areas A portion of a Forest larger than 5000 acres, or smaller than this but contiguous to designated wilderness or primitive areas, which contains no roads and has been inventoried by the Forest Service for possible inclusion in the wilderness areas system. By management directive these areas were not to be developed until a determination was made on an individual basis whether the area should be considered as a possible wilderness area.

Roads, Types *Permanent Roads* - Those system roads constructed for long term or periodic use. The three categories of permanent roads are: Constant Service - All Weather - Roads constructed to permit year-round hauling and travel, except during wet periods when closures may be necessary. These roads normally require surfacing and permanent drainage structures. Intermittent Service - Roads used for short periods of time for a single project and put to bed between use periods. These roads require little or no surfacing. Drainage structures are normally temporary and are removed between use periods. Winter Service - Roads on which use is limited to frozen ground conditions.

Temporary Roads - A nonsystem road constructed for a single project, usually for timber harvest, which has no projected use within 20 years of completion of the project. These roads require little or no surfacing. Drainage structures are temporary and are removed upon completion of the project. The roadway is returned to resource production.

Rock, Igneous Rocks formed by solidification from a molten or partially molten state, e.g., various types of lava. Igneous rock is often subdivided, for convenience, into "plutonic rocks" (those which were formed by the solidification of molten materials below the ground surface – e.g., granite) and "volcanic rocks" (those formed by solidification of molten materials which have been extruded onto the ground surface.).

Rock, Metamorphic Rocks which have been formed in the solid state under the conditions of high pressure, high temperature and the introduction of new chemical substances that, in general occur at great depths within the earth, e.g., slate, marble, and jade.

Rock, Sedimentary Rocks which have been formed from deposits of sediment, whether from fragments of other rock transported from their sources and deposited by water, e g., sandstone or shale — or by precipitation from solution or fixation by organisms, e.g., rock salt, gypsum, and limestone.

Rotation Process of the Earth turning on its axis. Rotation determines day and night, and the length of the day. Compare with revolution.

Rotation (Cutting Rotation) The planned period of years between initially establishing a stand of timber and the time when it is considered ready for cutting and regeneration.

Rotation Cycle The planned number of years between the regeneration of a stand and its final cutting at a specified stage of maturity.

Rotational Deferment A grazing system in which one or more parts of the range are rested during the growing season each year; and rotational use of other segments of the range are not necessarily planned.

Rough Reduction Burns Controlled ground fires used in pine types to reduce fire hazards caused by excessive flammable vegetation buildup.

Roundwood Trees that are used without being milled (fence posts, telephone poles, pulpwood, etc.).

RUA (Restricted Use Area) An area that is closed to specific activities to protect damaged or fragile areas and to give them a chance to recover, and to encourage the dispersal of hikers and campers. RUA's are applied only when and where needed.

Runoff 1. The portion of rainfall that is not absorbed by the deep strata and not lost to vegetation or evaporation. It is that which finds its way into streams as surface flow. 2. The total stream discharge of water (quantity or rate), including both surface and sub surface flow, usually expressed in acre feet or in cubic feet per second per square mile of drainage area.

S

Sacrifice Area A relatively small area of land irrespective of site in a grazing unit that may still be over-used after practical measures for securing uniform grazing distribution have been installed.

Sacrifice Site A range site that is intentionally overgrazed to obtain efficient overall use of the management area.

Sale Schedule The quantity of timber planned for sale by time period, from the area of suitable land covered by a Forest plan. The first period, usually a decade, of the selected sale schedule provides the allowable sale quantity. Future periods are shown to establish that long-term sustained yield will be achieved and maintained.

Sale-area Betterment Improvements in a forest that can be funded by K-V deposits.

Salvage Cutting Cutting primarily to utilize dead, deteriorating, and downed material and scattered poor-risk trees that will not be merchantable if left in the stand until the next scheduled cut.

Salvage Dead or dying trees that occur in excess of those needed for wildlife, aesthetics, or other purposes. These trees are harvested for Production.

Salvage Harvest of Timber The removal of dead trees or trees being damaged or killed by injurious agents other than competition, to recover value that would otherwise be lost.

Sample, Random A sample taken without bias from an area or from a population in which every part of the area or population has an equal chance of being taken.

Sampling The process of obtaining a sequence of discrete digital values from a continuous sequence of analogue data.

Sand Inorganic particles between 0.05 (and 1/16) and 2.0 millimeters in diameter. Also soil that contains 85 percent or more sand and a percentage of silt plus 1.5 times the percentage of clay that does not exceed 15.

Sanitary Landfill A site where solid waste materials are disposed of on land, supposedly in a manner that prevents their escape into — or pollution of-the surrounding environment. The waste is spread in layers, then compacted to the smallest practical volume and covered with compacted soil at the end of each working day.

Sanitation Cutting (Salvage) Sanitation cutting is removing and exploiting dead, dying, deteriorating, or susceptible trees to prevent the spread of pests or pathogens and so promote forest hygiene. The recovery of trees materially damaged by fire, wind, insects, fungi, or other injurious agencies before their timber becomes commercially valueless.

Sanitation Harvest of Timber The removal of trees to improve-stand health and to reduce actual or anticipated spread of insects and disease.

Sapling A loose term for a young tree greater than a few feet tall and an inch or so in diameter at breast height and typically growing vigorously and without dead bark or more than an occasional dead branch. Many countries set arbitrary size limits e.g., 2 to 4 inches diameter at breast height. As used in timber survey, a size-class definition; trees 1.0 to 4.9 inches at DBH.

Satellite A free-flying object that orbits the Earth, another planet, or the sun.

Satellite Revolution The time from one perigee (the point of an elliptical orbit path where a satellite is closest to Earth) to the next.

Satisficing A method of choosing among alternatives. Minimum values or quality standards are established for all attributes affected by the alternatives. All those alternatives that do not meet the minimum requirements are eliminated from consideration. By successively raising the minimum requirements you can successively reduce the number of alternatives under consideration.

Sawtimber As used in timber survey, a size class definition, trees larger than 9 inches at dbh.

S-Band One of the segments or bands into which the radio frequency spectrum above 1000 MHz is divided, designated by letters. Signals from GOES and other geostationary spacecraft transmitting on or near 1691 MHz are transmitting on S-Band.

Scale The proportional relationship (ratio) between the reduced size at which something is being represented on a map or other type of drawing and its true distance or size relationships. In aesthetics, a feeling for the size and/or appropriateness of the size of some construction element (i.e., a building, paved area, etc.) or project.

Scaling In forestry, measuring the volume of logs.

Scanner A system that optically scans its detector(s) across a scene and records or stores the data in a two-dimensional format to form an image.

Scanning radiometer An imaging system consisting of lenses, moving mirrors, and solid-state image sensors used to obtain observations of the Earth and its atmosphere. Scanning radiometers, which are the sole imaging systems on all current operational weather satellites, have far better long4erm performance than the vidicon IV camera tubes used with earlier spacecraft.

Scarification Loosening the top soil in open areas to prepare for regeneration by direct seeding or natural seed fall.

Scattering The process by which electromagnetic radiation interacts with and is redirected by the molecules of the atmosphere, ocean, or land surface. The term is frequently applied to the interaction of the atmosphere on sunlight, which causes the sky to appear blue (since light near the blue end of the spectrum is scattered much more than light near the red end).

Scenario A word picture of an hypothetical sequence of future events constructed for the purpose of focusing attention on causal processes and decision points. "Scenarios" may express a series of integrated objectives for preserving the environment, the fulfillment of which is dependent upon timing as an "urgency" rather than as a "sequence". For example, some strategic natural areas that may be shown on scenarios actually exist on the ground today, but will not be found anywhere in the year 2000 unless they are reserved now.

Scenery Management System An overall framework for the orderly inventory, analysis, and management of scenery.

Scenic Area A place designated (such as cliffs, streams, vistas, vegetation, and wildlife) as containing outstanding or matchless beauty which requires special management to preserve these qualities.

Scenic Corridor The visible land area outside the highway right-of-way and generally described as "the view from the road".

Scenic River Rivers or sections of rivers that are free of impoundments, with shorelines or watersheds still largely primitive and shorelines largely undeveloped, but accessible in places by roads.

Scenic Vista A point along a route of travel that affords a panoramic, unusual, or highly-pleasant view.

Schema Entry All the data at all levels stored for one of the major items in a database.

Schema Family Those schema records vertically related as one single branch of the schema.

Schema Record A group of components related to the same subject that may have multiple occurrences of data.

Schema The complete list of all data base component numbers and names, and their Key or Non-key designation, item type, and field length. Also includes all Strings and Stored Functions.

Science Consistency Review A process to determine whether scientific information of appropriate content, rigor, and applicability has been considered, evaluated, and synthesized in the documents that underlie the land management plan approval. The *science consistency review* does not advise the decisionmaker for or against a particular course of action.

Science Science is the knowledge, information, concepts, and theories based on organized systems of facts that have been learned from study, observation, and experience.

Scope and Scale *Scope* usually refers to the breadth, extent, or depth of an issue, such as whether it refers to one or multiple resources, ecosystems, or species. *Scale* usually refers to geographic scale, such as an ecosystem, landscape, or eco-region.

Screaming eagles Cloud pattern so named because some observers maintain they can see the head of an eagle facing west in these cloud patterns. The pattern is similar to a comma, only the pattern is disorganized and not solid. Weather associated with screaming eagles consists of rain showers and gusty surface winds up to about 25 knots. The eagles can intensify and enlarge when moving into areas east of troughs; in that case, intense thunder storms can develop. Screaming eagles are common in the Pacific Ocean between Hawaii and the equator and are uncommon in the western Atlantic.

Screen Record Information about openings that permit water to enter a well, including perforations and uncased sections of the aquifer.

Screening Topographic or vegetative features that block, cover, or obscure views of human activity, past or present, or of other persons.

Sea breeze Local coastal wind that blows from the ocean to land. Sea breezes usually occur during the day because the heating differences of land and sea cause pressure differences. Cooler heavier air from the sea moves in to replace rising warm air on the coastline. See land breeze.

Sea level The datum against which land elevation and sea depth are measured. Mean sea level is the average of high and low tides.

Search and Rescue International satellite-aided search and rescue project. COSPASI SARSAT satellites monitor the entire surface of the Earth, and transmit distress signals to special ground receiving stations. The receiving stations compute the location of the signal, and notifv the nearest rescue coordination center. Satellite search has cut recovery time from days to hours, and has aided downed airplanes, capsized boats, and persons in other emergencies.

Seasonal Use Road Facilities developed and operated for only seasonal use because of climatic, structural or administrative limitations.

Second Growth Forest growth that has come up naturally after some drastic interference (e.g., wholesale cutting, serious fire, or insect attack) with the previous forest crop.

Secondary Productivity The rate at which organic matter is produced by the heterotrophic organisms of a community.

Secondary Range Range that is lightly used or unused by livestock under minimal management and will ordinarily not be fully used until the primary range has been overused.

Secondary Use (Accessory Use) A use that is incidental, related, appropriate, and clearly subordinate to the main use of an area, lot, or building. An accessory use does not alter the principal use or properties of the area.

Secondary Wastewater Treatment Any process that takes the effluent from primary wastewater treatment and reduces the suspended solids and biological oxygen demand by approximately 90 percent. During secondary treatment, the wastewater is subject to biochemical action, increased sedimentation, and clarification.

Security The protection inherent in any situation that allows animal to remain in a defined area despite an increase in stress or disturbance associated with the hunting season or other human activities. A state of being or a condition. The components of security may include, but are not limited to, vegetation, topography, area, road density, distance from roads, size of vegetation blocks, hunter density, season timing, and land ownership. Hiding cover is site specific, while security is area specific.

Security Area Any area that will hold animal during periods of stress because of geography, topography, vegetation, or a combination of those features, more meaningful than security habitat.

Sediment Solid materials, both mineral and organic, that is in suspension, is being transported, or has been moved from its site of origin by air, water, gravity, or ice that has come to rest on the Earth's surface either above or below sea level.

Sediment Delivery Ratio The fraction of the soil eroded from upland sources that actually reaches a continuous stream channel or storage reservoir.

Sediment Discharge The quantity of sediment, measured in dry weight or by volume, transported through a stream cross section in a given time. Sediment discharge consists of both suspended load and bedload.

Sediment Yield The quantity of sediment, measured in dry weight or by volume, transported through a stream cross-section in a given time.

Seed Tree A tree purposely left standing at the time of cutting in a forest, for the purpose of producing seed for regeneration of trees in the cutover area.

Seed Tree Cutting (Seed Cutting) Removal in one cut of the mature timber from an area, except for a small number of seed bearers left singly or in small groups. A regeneration cutting where the planned source of an adequate new stand is from seed existing on, or to be produced by, trees standing on the cut-over area.

Seed Tree Removal Cutting Cutting made to remove the seed trees when adequate reproduction above browse height has been established.

Seed, Certified Seed for a named variety of plant (blue tag). Seed of a plant species for which there are no released varieties (or there are ecotypes of interest that are distinctly different from any released varieties). These may be obtained as "Certified." Foundation Seed (white tag) and Registered Seed (purple tag) - Seed used by seed growers to produce certified seed. It has been field inspected and laboratory tested to ensure varietal identity, genetic purity, high seed purity and germination, and limited amounts of other crop seeds, weed seed, inert matter, and diseased or insect-damaged seed. Source Identified Class - This is also "certified" seed. It is seed for which there has been verification by a see certification agency about the state, county, and elevation from which the seed lot was collected (yellow tag). Selected Class - Seed of an ecotype or germplasm that has been compared with other ecotypes or germplasm selection of a species and shown to have unique characteristics (green tag). Tested Class - Seed of an ecotype or germplasm selection that has been selected and also progeny tested to prove that unique traits are heritable (blue tag).

Seedbed In natural regeneration, the soil or forest floor in which seed falls. In nursery practices, prepared area that is seeded.

Seedling As used in timber survey, a size class definition, trees less than one inch at DBH.

Seismic Hazards Hazards related to earthquake activity.

Seismic Pertaining to or caused by an earthquake.

Selection Cut The periodic removal of mature trees, individually or in small groups from an uneven-aged forest. By this method both regeneration cutting and tending of immature stand components are accomplished at each entry.

Selection Harvest Cut A system that removes trees individually in a scattered pattern from a large area each year. 1. Individual tree selection cutting involves the removal of selected trees of all size classes on an individual basis. Regeneration is established under the partial shade of the overstory canopy after each cut.

2. Group selection cutting involves the removal of selected trees of all size classes and groups of a fraction of an acre of to 2 to 3 acres in size. Regeneration occurs in the groups under conditions similar to those found in small clearcuts.

Selection Logging (Selection Cutting) Regeneration cut designed to create and perpetuate an uneven-aged stand. Removal of mature timber, usually the oldest or largest trees, either as single scattered trees or small groups at relatively short intervals, commonly 5 to 20 years, repeated indefinitely, by means of which the continuous establishment of natural reproduction is encouraged and an uneven-aged stand is maintained.

Selection System An uneven-aged silvicultural system in which trees are removed individually, here and there, from a large area each year. Regeneration is mainly natural and the timber crop is ideally of all different ages.

Selective Herbicide An herbicide which is effective only against centain species and is able to control unwanted plants without serious injury to desirable species.

Selectivity The degree to which a plant or plant part is removed differently from that which is expected by random removal. Selectivity is expressed as the ratio of the proportion of a plant or plant part in a diet to all the other herbage.

Self-Service Level A level of recreation operations, administration, and maintenance under which mandated health and safety tasks are performed. Visitors are expected to help maintain areas in a clean condition and to serve themselves.

Self-Sustaining Populations Populations that are sufficiently abundant, interacting, and well-distributed in the plan area, within the bounds of their life history and distribution of the species and the capability of the landscape, to provide for their long-term persistence, resilience, and adaptability over multiple generations.

Semi major axis One of the six Keplerian elements, it indicates the size of an orbit. The semi-major axis is one half of the longest diameter of an orbital ellipse, e.g., one-half of the distance between the apogee and perigee of an Earth orbit. (The semi-major axis is related to the orbital period and mean motion by Kepler's third law.

Semi-Primitive Motorized Class A classification of the recreation opportunity spectrum characterized by a predominantly unmodified natural environment in a location that provides good to moderate isolation from sights and sounds of people, but allows use of motorized equipment.

Semi-Primitive Non-Motorized Class A classification of the recreation opportunity spectrum characterized by a predominately unmodified natural environment of a size and location that provides good to moderate opportunity for isolation from sights and sounds of people. No motorized equipment is allowed..

Sense of Scale A feeling for the size and/or appropriateness of the size of some project or construction element (i.e., building, paved area, etc.). An evaluation of the sense of appropriate of the size (or mass) relationship of some design material, element, or object to its surroundings, its intended use or user group, etc. For example, the use of brick as a building material is sometimes said to result in constructions of a "human scale"—because people have a strongly developed awareness of the size of individual bricks, and therefore, can sense the size of buildings or paved areas by using the individual brick as the basis of judgement. In contrast, because concrete can be poured into almost any size units, people can have difficulty correctly judging the size of objects made out of concrete. In a similar sense, entire large high-rise (i.e., multi-storied) buildings may be described as being out-of-scale when placed in surroundings dominated by 2- or 3-story buildings.

Sensitive Areas Areas of high erosion hazard, areas that may be susceptible to compaction, or areas of unstable slopes.

Sensitive Species Species identified by criteria below what are known, reported, or suspected to occur on or in the immediate vicinity. The criteria are: 1. Species undergoing placement on an official list;. 2. Species placed on an agency list because they require special management attention. Examples: (a) Species common elsewhere, but a disjunct population of unique, popular, or scientific interest. (b) Locally endemic population in unique habitats that warrant continued monitoring or special management to assure jeopardy is not occurring and will not occur in the future.

Sensitivity Analysis Determination the consequences of varying the level of one or more of several factors while holding other factors constant. Varying the value of a parameter to find the extent to which the change affect the results of the analysis. If a small change in an assumption results in a proportionately greater change in the results, then the results are said to be "sensitive" to that assumption or parameter.

Sensitivity Level As used in Visual Quality Management; a particular degree or measure of viewer interest in the scenic qualities of the landscape. 1 - most sensitive, 2- sensitive, and 3 - less sensitive; an index to beauty based on criteria having factors affecting landscape aesthetics.

Sensitivity, Land The degree to which the land cannot sustain or heal itself from externally imposed disturbance or land uses.

Sensor Device that produces an output (usually electrical) in response to stimulus such as incident radiation Sensors aboard satellites obtain information about features and objects on Earth by detecting radiation reflected or emitted in different bands of the electromagnetic spectrum. Analyzing the transmitted data provides valuable scientific information about Earth.

Weather satellites commonly carry radiometers, which measure radiation from snow ice, clouds, and bodies of water Spaceborne radars are used for Earth observations, bouncing radar waves off land and ocean surfaces to study sea-surface conditions, ice thickness, and land surface features. A wind scatterometer is a special type of radar designed to measure ocean surface winds indirectly by boun& mg signals off the water and measuring them from various angles. Infrared (IR) detectors measure heat generated by Earth features in the IR band of the spectrum.

Photographic reconnaissance sensors in their simplest form are large telescope-camera systems used to view objects on Earths' surface. The bigger the lens, the smaller the object that can be detected. Camera-telescope systems now incorporate all sorts of Sophisticated electronics to produce better images, but even these systems need cloudless skies, excellent lighting, and good colour contrast between objects and their surroundings to detect objects the size of a basketball. Some of the satellites produce film images that must be returned to Earth, but a more convenient method is to record the image as a series of digital code numbers, then reconstruct the image from the electronic code using a computer at a ground station.

Sensor calibration The relationship between input and output for a given measurement.

Seral Of or about the series of stages in ecological or plant succession.

Seral Habitat A biotic community that is a developmental, transitory stage in an ecological succession. <>.

Seral Stage A recognizable step or stage in the development of ecological communities. Developmental temporary communities in a sere; not fixed.

Sere A developmental series of communities a chain of seral stages containing the initial (pioneer), one or more transitional stages, and a single (often hypothetical) climax stage.

Servient The land against which an easement or privilege exists is called the "servient" tenement and the estate to which it is annexed is the dominant tenement. Their owners are respectively the servient and dominant owners.

Shade Intolerant Those tree species that need full or near full sunlight to regenerate and grow.

Shade Strip A strip of vegetation along the edge of surface waters requiring special management to protect the surface waters from adverse impacts from solar radiation.

Shade Tolerant Relative ability of a tree species to reproduce and grow under shade. Tree species are usually classified in descending order of shade tolerance as: very tolerant, tolerant, intermediate, intolerant, and very intolerant.

Sheet Erosion Removal of a fairly uniform layer of soil from the land surface by runoff water, without the development of conspicuous water channels.

Shelterbelt A wind barrier of living trees and shrubs maintained for the purpose of protecting farm fields. As applied to individual farmsteads, termed a windbreak.

Shelterwood Cut Removal of the mature timber in a series of cutting over a period of years to encourage natural reproduction under a partial cover of seed trees. About half-shade or more on the ground is retained; grass and brush competition is thereby suppressed.

Shelterwood Cutting A cutting method used in even-aged management. It is the removal of a stand of trees through a series of cuttings designed to establish a new crop with seed and protection provided by a portion of the stands.

Shelterwood Method A regeneration method under an even-aged silvicultural system under which a portion of the mature stand is retained as a source of seed or protection during the period of regeneration. The mature stand is removed in two or more cuttings commonly termed seed cutting and removal cutting. The seed cutting may or may not be preceded by a preparatory cutting.

Shelterwood System - Even-aged silvicultural systems in which (in order to provide a source of seed and/or protection for regeneration) the old crop is removed in two or more successive "shelterwood cuttings", the first of which is ordinarily the seed tree cutting and the last is the final cutting, any intervening cuttings being termed removal cuttings.

Short-range Planning Planning for a future 5 years or less distant.

Short-Term Effects Those effects that will usually occur within the 20 years after an action.

Short-Term Facility (Road) A facility developed and operated for a limited period of time which will cease to exist as a transportation facility after the purpose for which it was constructed is completed, and the occupied land is reclaimed and managed for natural resource purposes.

Shrink-Swell Clays Clays that swell when moisture increases, and shrink when moisture decreases.

Shrink-Swell Potential The probability and extent of a soil's volume to change due to loss or gain in moisture content.

Shrub A low-growing perennial plant with a persistent woody stem and low branching habit.

Sight Distance The distance at which 90 percent or more of an adult big game animal is hidden from human view. Measures of the effectiveness of hiding cover.

Signal Electrical impulses, sound or picture elements, etc., received or transmitted. Signals can exist in many different forms and media (electrical/wires, acoustic/air light/transparent fibres, etc.), but all signals will vary with time.

The signal shape plotted as a function of time is called the waveshape or waveform. Some waveforms are repetitive or periodic, that is, a small segment of the waveform repeats itself regularly. Other waveforms, such as noise, are nonperiodic or aperiodic. All waveforms can be distilled into the combination of pure waves called sine waves. The frequency of a sine wave is the rate at which the fundamental shape repeats itself.

Most signals occupy a limited range of frequencies between a lower limit and an upper limit. This range or band of frequencies occupied by a signal is called the bandwidth of the signal.

Communication medium or channel can pass only a specific range or band of frequencies, which is called the bandwidth of the channel. The bandwidths of the channel and the signal determine the number and types of signals that can be transmitted by a particular communication channel. Signals often are too small and need to be made larger through a process called amplification. The amount of amplification is measured in decibels. However amplification is an imperfect process, and inadvertently introduces various distortions, noise, and bandwidth limitations. Often, multiple signals must share the same medium. One way the sharing can be accomplished is to place each signal in its own band of frequencies within the total band of the medium. The combining of a number of signals to share a medium by dividing it into different frequency bands for each signal is called frequency-division multiplexing.

Frequency-division multiplexing requires the ability to move signals around so that each multiplexed signal occupies its own band. This is accomplished through a process called modulation, in which a high-frequency sine wave carries the signal into the specified band. Either the amplitude or the frequency of the carrier wave can be varied, or modulated, in synchrony with the information-bearing signal. These methods are called amplitude modulation (AM) and frequency modulation (FM). FM is the more complex process of the two, and the bandwidth of the FM carrier can be many times that of the modulating signal. The process of demodulating a frequency-modulated signal eliminates much of the deleterious effects of additional noise. (The trade-off between bandwidth and noise immunity characterizes most communication systems. Both are analogue modulation schemes for multiplexing signals in the frequency spectrum.).

Digitizing a signal requires a number of steps and results in a binary digital signal that takes on one of two discrete values. This process results in considerable immunity to additive noise, but requires a considerable increase in bandwidth.

Signal to noise ratio (SNR) In decibels (dB), the difference between the amplitude of a desired radio frequency (RF) signal and the internal or external RF noise level in a system. A negative SNR indicates the signal is below the system noise level and unusable. The greater the positive SNR, the less effect noise will have on the final quality. SNR of at least + 12dB is necessary to produce imagery with minimal noise effects.

Significance, Statistical An effect is said to be *statistically significant* if the value of the statistic used to test it lies outside acceptable limits, that is to say, if the hypothesis that the effect is not present is rejected. A test of significance is one which, by using a test statistic, purports to provide a test of the hypothesis that the effect is absent. By extension, the critical values of the statistics are themselves called significant.

Silt Fragments of rock or organic matter sediment most of the individual particles of which are between 1/16 and 1/256 millimeters (0.05 and 0.002 millimeters) in diameter. A soil texture type consisting of 80 percent or more of silt particles and less than 12 percent clay particles.

Silt Trap A physical barrier installed between a disturbed area and an area to be protected that filters suspended sediment from surface runoff.

Silvics The science which deals with the laws underlying the growth and development of single trees and of the forest as a biological unit.

Silvicultural Prescription A statement following an examination of a compartment usually done at least once every 10 years to determine management needs of the timber. It includes actions to be taken based on soils, wildlife needs, silvics of each species, ownership patterns, engineering needs, and goals and objectives of the organization.

Silvicultural System A management process whereby forests are tended, harvested, and replaced, resulting in a forest of distinctive form. Systems are classified according to the method of carrying out the fellings that remove the mature crop and provide for regeneration and according to the type of forest thereby produced.

Silviculture Generally, the science and art of cultivating (i.e., growing and tending) forest crops. More particularly, the theory and practice of controlling the establishment, composition, constitution, and growth of forests. The application of the knowledge of silvics in treating a forest;.

Simulation 1. The imitative representation of the functioning of one system or process by means of the functioning of another. A mathematical or computer model of an ecosystem. Examinations of a problem, often not subject to direct experimentation, by means of a simulating device.

2. A technique for solving complex problems that is not amenable to solution using formal analytical techniques. Essentially simulation consists of a representation of a system or organization by means of a model. The behaviour of the system under various possible operational conditions or assumptions is then analyzed through repeated manipulation of the model.

Sine wave A smoothly varying wave that repeats itself; its frequency is the rate at which the fundamental shape repeats itself. Any waveform can be distilled into a combination of pure sine waves of varying frequencies and amplitudes.

Single Tree Selection Cutting The cutting method that describes the silvicultural system in which trees are removed individually, here and there, over an entire forest or stand. The resultant stand usually regenerates naturally and becomes unevenaged.

Sink The process of providing storage for a substance. For example, plants – through photosynthesis – transform carbon dioxide in the air into organic matter which either stays in the plants or is stored in the soils. The plants are a sink for carbon dioxide.

Sinkhole A depression in the landscape formed by the collapse of the land surface below which limestone has been dissolved. Karst topography has many sinkholes.

Site A plot of land intended or suitable for development or an area considered in terms of its environment (biological, climatic and soil factors present), particularly as this determines the type and quality of the vegetation the area can inherently produce. "Sites" are classified either qualitatively, by their climate, soil and vegetation, into site types, or quantitatively, by their potential wood production, into site classes.

Site Capacity The number of persons who may, at one time, use the developed capacity (i.e., facility carrying capacity) of a location while not exceeding the safe, season long carrying capacity (i.e., ecological carrying capacity) per acre.

Site Class A quantitative measure of the productive capacity of an area which is essentially uniform with respect to those factors controlling productivity for the crop or stand understudy. The quantitative measure is based on the volume, height, or maximum mean annual increment of dominant types that is attained or attainable on that area at a given crop age.

Site Index A numerical evaluation of the quality of land for plant productivity, especially used in forest land where it is determined by the height on one or more of the tree species, usually at 50 years.

Site Plan A plan, prepared to scale, showing accurately and with complete dimensioning, all of the buildings, structures and uses and the exact manner of development proposed for a specific parcel of land.

Site Planning (Site Design) Organizing the external physical environment up to the largest scale at which it can still be subject to unified and complete control. It deals with structures, land and the entire complex of physical forms above, below and on the surface. It has at its heart the disposition of objects and activities in 3-dimensional space. It begins with the careful analysis of the project site and purpose; and, it concludes with design of a pattern which interacts as a totality with its users, and which is subject to continuous future development and change. "Site planning" produces plans that can be carried out in one continuous foreseeable process, according to one original design, under the control of one agency, inclusive of all details of site engineering, landscaping, and architecture. "Site planning" may take place on areas as small as a single building and its grounds, or as large as the layout of a complete small town.

Site Preparation Preparing a regenerated stand by hand or mechanical methods or prescribed burns or herbicides to stimulate the establishment and growth of a new timber stand. Purpose is to control vegetation that would compete with new growth. It involves the removal of residual trees left after a commercial harvest and preparation of seedbeds. Mechanical methods include use of a crawler tractor with a KG blade that shears vegetation near ground level. Hand methods include use of ax and chainsaw, or tree injectors.

Site Specific Something only valid for, or confined to, a certain given parcel of land and/or water. The term may apply to data, studies to obtain information, environmental impacts, use restrictions, etc. The assumption is that mapped land characteristics are, in varying degrees, generalizations about a true distribution of characteristics. Such generalizations cannot be assumed valid for every parcel of land contained within the larger mapped units. "Site specific" studies may be needed to confirm or change the appropriate land use classification of a particular parcel.

Site Type A biotic productivity unit (usually for trees) which is qualitatively defined on the basis of being relatively uniform with respect to climate, soil and natural vegetation cover.

Skid Road or Trail Travelway used to drag or transport trees from the stump to a landing.

Skidding A loosely used term for transporting logs from stumps to a collecting point by sliding or dragging along the ground-as opposed to the use of wheels, helicopters, balloons, cables, etc., to keep them totally off the ground. The logs may slide more or less wholly along the ground or with their forward end supported. The method of skidding can greatly affect the impact of logging on soil and the residual stand.

Skylab The first U.S. space station, launched unmanned in May 1973 and soon after occupied in succession by three crews through November 1973.

Slab The exterior portion of a log removed during the sawing process.

Slash The residue or debris left on the ground after timber cutting, pruning, thinning, or brush cutting and/or accumulating there as a result of storm, fire or other damage. It includes unutilized logs, up-rooted stumps, broken or uprooted stems, branches, twigs, leaves, bark, and chips.

Slash Disposal Treatment of slash to reduce the fire hazard or for other purposes.

Slip A mass of spoil material that moves downward and outward to a lower elevation due to the force of gravity, generally caused by over loading of the downslope, freezing and thawing, or saturation of the fill.

Slope (Grade) The inclination of a stream channel or ground surface, usually expressed in terms of the ratio or percentage of number of units of vertical rise or fall per unit of horizontal distance. An increase of 1 foot over a distance of 5 feet is expressed as a 20 percent slope. The slope of a terrestrial surface is an important factor in erosion potential and interacts with aspect to influence basic rates of evapotranspiration.

Slope Orientation (Aspect, Exposure) The compass direction toward which the slope of a land surface faces (e.g., north, and northwest)(the direction downhill). Giles et al. recommended 2 *types* of orientation, those N/S and those East/West, since each has very different ecological influences.

Slope Stability (Mass Instability) An evaluation (almost always qualitative and expressed as a probability) of the tendency for the materials on or constituting a slope (e.g., rocks, soil, snow) to either remain in place or to move downhill.

Slope Stability The resistance of any inclined surface, as the wall of an open or cut, to failure by sliding or collapsing.

Slope Steepness (Slope, Slope Gradient, Slope Inclination) The degree of deviation of a land surface from the horizontal usually expressed in percent (i.e., amount of elevation change per 100 feet, expressed as a percentage) or degrees. A 45-degree slope is a 100 percent slope.

Slurry Refuse separated from the coal in the coal cleaning process of relatively small size which is readily pumpable in the washing plant effluent. A pulverized coal-liquid mixture transported pipeline.

Snag A standing dead tree (20 or more feet tall and over 4 inches DBH) potentially or actually used by birds and other animals for nesting, roosting, perching, breeding and/or foraging for food.

Snow Avalanche Hazard Some evaluation (almost always qualitative and relative to adjacent locations) of the probability that a snow avalanche will occur on a given site.

Snow Avalanche The rapid downslope movement of large quantities of snow.

Sociability The distribution of organisms in relation to one another as individuals or as groups within a community. J. Braun-Blanquet recognizes five classes of sociability, ranging from isolated individuals to dense masses. Social activity marked by a feeling of unity but also by individual participation completely willing and not forced or coerced and without a loss of individuality.

Social Analysis An analysis of the social (as distinct from the economic and environmental) effects of a given plan or proposal for action. Social analysis includes identifying and evaluating of all pertinent desirable and undesirable consequences to all segments of society, stated in some comparable quantitative terms such as persons or per cent of population in each affected social segment. It also includes a subjective analysis of social factors not expressible in quantitative terms. It may include effects on life styles, the political process, education, health care, housing, distributions of power and income, and the quality of public services. Procedures include: questionnaire design, distribution and retrieval; conducting polls; holding public hearings; forming ad hoc task forces and advisory groups; and conducting behavioural studies.

Social and Economic Elements The variety of tangible and intangible uses, values, products, services, opportunities, and benefits provided by National Forest System lands.

Social Benefits The net benefit considerations of long-range societal values at the regional or national level that might not be taken into account in the profit and loss statement of an individual farmer, forest operator, industrialist or other private citizen. Exclusive of corporations, profits and losses appear in the statement of single individuals. Also, the non-monetary and rarely quantifiable returns to society arising from any form of economic activity e.g., those recreational benefits resulting from the creation of a scenic over look.

Social Class People who have been grouped together on the basis of one or more common characteristics. The characteristics can usually be viewed as hierarchical.

Social Costs The net losses of long-range societal values at the regional or national level that might not be taken into account in the profit and loss statement of an individual farmer, forest operator, industrialist or other private citizen. Note: individual not corporate emphasis. These are the non-monetary and rarely quantifiable losses to society arising from any form of economic activity– e.g., from air pollution whose only deleterious side effect is reduced visibility, over and above the expenditures for the goods and services causing the smoke.

Social Determinism (Social Primacy) A point of view in planning that advocates that the primary basis for making use and allocation decisions should be satisfaction of current societal demands on resources. The ability of the resource base to sustain such demanded use levels over the longrun is not recognized as an important criterion for decision making by the social primist. Thus is grounded in the belief that human behaviour is produced by conditions in the social world. Social Determinism (continued) - In this view, when a person is born, he or she is essentially unmarked, a clean surface. Potentials for good or evil, success or failure, etc. will be decided by the social environment, which is formed by family, friends, work associates, and overlaying all of these groups, the culture.

Social Distance The physical distance apart maintained by animals and persons based on the social situation and culture. Also, people are conscious of having a certain "place" in a social hierarchy and, they are conscious of the places of others. The differences between the respective places are "social distance". For example, the perceived social distance between a physician and a garbage collector will probably be greater than that between the physician and a lawyer. Occupational status is only one component of "social distance"; others will be relevant to the situation in which the individuals find them selves. For example, on a ski slope, occupation may mean little and skiing ability may be the essential determinant of social distance. Social distance will also vary with the extent to which the two people know each other. On first encounter the determinants of social distance will include appearance, mode of self-introduction, and attributes of social status such as occupational prestige. As the relationship develops, these factors may become less important, and others, such as intelligence, will enter into determining social distance.

Social Dynamics The changes in people resulting from forces and processes. The processes of interaction leading to change may involve economic, political, ideological, individual, or other variables. Some theorists see this dynamic as a dialectical one, involving a succession of conflicts and their resolutions. Others see the dynamic as a developmental, cumulative process in which conflict is merely incidental.

Social Equity The distribution of the gains and losses that will accrue to individuals or groups (defined according to social criteria), as a consequence of land use planning decisions, in a manner which is in reasonable conformity to accepted standards of natural rights, law and justice, without prejudice, favouritism or fraud and without causing undue hardship.

Social Inequality A condition in which various members of a society have unequal amounts of income, prestige, and social power.

Social Mobility The upward or downward movement of individuals or groups into different positions in a social hierarchy based on wealth, income, occupation, education, social power, or any other scarce social resources. A society with a high rate of social mobility and good opportunities for upward social mobility is not egalitarian but may maximize opportunities for one individual or group to get more of society's scarce resources than other individuals or groups can get.

Social Mores The fixed customs or folkways of a particular social group that are morally binding upon all members of the group and seem necessary to its welfare and preservation.

Social Norms The shared expectations of a social group or society. The concept assumes a relationship between the perceptions of shared expectations by members of the social group and the extent to which the norm influences behaviour. To the degree to which human behaviour is influenced by the normative content of the culture of one's society, such behaviour might be explained in terms of the collective expectations as perceived, shared, and enforced by the group. Normative variations exist even in a relatively stable culture. These variations are due to the generalized nature of norms relative to specific social situations and to the variations in their perception and interpretation by individuals. Moreover, in societies, where structure manifest subcultural variations, ambiguities and variations concerning the meaning of norms are to be expected. Norms are external to the individual to the extent that normative definitions exist in the reality defined by a particular culture. Such cultures may preexist his or her being. There is a hierarchy of such definitions of "right and wrong". These are traditionally identified as "folkways", mores, and laws.

Social Power The extent to which a person can impose their will on a group. That is a person has power when he gets the members of a group to do what he or she wants, regardless of what they want.

Social Prestige The respect or favourable regard enjoyed by individuals or groups and linked to the social status accorded to these individuals or groups. Prestige is frequently, but not necessarily, the result of the possession of power or wealth.

Social Role Generally, a role consists of those behaviours typically performed by an individual in a particular situation. The individual's assumption of a role implies his or her acceptance of the need to act in ways that are socially agreed upon as being appropriate in that situation. A status generally refers to an institutionalized role. Thus, statuses often involve occupational or kinship systems, e.g., physician, lawyer, factory worker, or father, mother, uncle. Whereas the physician, by virtue of his or her profession, enjoys a certain "status", his or her "role" in any situation will be more fluid.

Social Status A position based on prestige and life style. Status group members share these characteristics to a similar degree. It is often useful to consider a society's status hierarchy independently from its social class hierarchy that is based largely on economic position. Status frequently derives from class position, but it may be achieved independently of economic standing.

Social Well-being One of the four required accounts categorizing, displaying, or "accounting" the beneficial and adverse effects of each alternative plan formulation for water and related land resources planning. The "social well-being" account includes (at least): 1. Real income distribution among individuals, classes, and groups;.

2. Life, health, and safety;.

3. Educational, cultural, and recreational opportunities; and.

4. Emergency preparedness.

Soft Mast Soft fleshy fruits potentially eaten by wildlife (examples: persimmon, wild grapes, blackberries, blueberries, huckleberries, mulberries, plums, autumn olive, and crabapples).

Software The programs, data, or routines used by a computer distinguished from the physical components (e.g., hardware).

Softwood A coniferous tree, one belonging to the botanical group gymnospermae; trees that in most cases have needle or scale-like leaves; also, the wood produced by such trees.

Soil 1. The loose top layer of the Earth, usually consisting of disintegrated rock with an admixture of organic matter and soluble salts, dynamic in nature, and serving as a natural medium for the growth of land plants. It differs from the material from which it is derived in many physical, chemical, biological, and morphological properties and characteristics. It usually only extends to the depth important for plant growth, the depth to root growth limiting material or 60 inch whichever is encountered first. Spatially, engineers recognize that there is an ungeneralizable, infinite variety of soils, each with slightly differing physical characteristics, the *regolith*. Agricultural soil scientist believes that "soils" can be mapped as significantly homogeneous, discrete units, i.e., soil types. It includes the surface soil (horizon A), the subsoil (horizon B) and the upper portion of the substratum (horizon C) to the extent that it is penetrated by plant and tree roots. The average soil is composed of 45% mineral, 25% air, 25%- water and 5% vegetation.

2. Surface layer of the Earth, ranging in thickness from a few inches to several feet, composed of finely divided rock debris mixed with decomposing vegetative and animal matter which is capable of supporting plant growth.

3. The mineral substrate plus the living and dead organic material within it. The biotope may be, in part, soil. Soil may be geochemically separated through time.

Soil and Water Mitigation Measures applied that significantly reduce the amount of potential soil erosion and sedimentation from resource management activities.

Soil Association A group of defined and named soil taxonomic units occurring together in an individual and characteristic pattern over a geographic region.

Soil Climate The moisture and temperature conditions existing within the soil. The soil has a microclimate that is the environment of the organisms that live within the soil.

Soil Compaction Reduction of soil pore space volume (porosity) which results in alteration of the soil chemical and physical properties. The "bulk density" increases with compaction.

Soil Complex A mapping unit used in detailed soil surveys where two or more defined taxonomic units are so intimately intermixed geographically that it is undesirable or impractical to separate them. The complex is a more intimate mixing of smaller areas of individual taxonomic units than the soil association.

Soil Condition Class A reference to soil stability based principally on the amount of ground cover weighted by the degree of accelerated erosion.

Soil Enhancement Adding lime and/or fertilizer to low productive soils in order to increase the production of market goods, plant health, and quality as forage.

Soil Fertility The quality of a soil that enables it to provide nutrients in adequate amounts and in proper balance for the growth of specified plants.

Soil Horizon A layer of soil that differs in composition, colour, or structure, or in both, from adjacent layers. A diagram representing a vertical section of the soil called a profile generally describes the various horizons in a soil. The horizons in a soil profile are designated as follows: A-horizon, topsoil; B-horizon, mineral soil; and C-horizon, parent soil material, weathered or unweathered rock fragments and minerals. Between any of the above horizons may be found a waterlogged or "gley" horizon, formed by the influence of ground water and marked by the presence of iron and other reduced compounds, alkaline salts, or aluminum oxides. A layer under the C-horizon that is unlike the parent material may be indicated as a D-horizon on a soil profile. The A, B, and C-horizons are quite distinct in mature soils, developed under forest vegetation in temperate and humid climates. These are called "zonal soils". Soils in which the horizons are undistinguished are called "azonal." Where the horizons do not fall into the above pattern or some horizons are absent, the soils are called "intrazonal.".

Soil Improvement Reshaping, liming, fertilizing, and revegetating of soils to improve the watershed condition and other renewable resources.

Soil Loss Equation (Universal Soil Loss Equation) A = RKLSPC wherein A = average annual soil loss in tons per acre per year, R= rainfall factor, K = soil erodibility factor, L = length of slope, S = percent of slope, P = conservation practice factor, and C = cropping and management factor. The equation as originally developed predicted soil losses from agricultural lands induced by rainfall. A recent revision takes into account the effect of snowmelt in areas where the contribution from this cause is significant.

Soil Loss Tolerance (Permissible Soil Loss, T Factor) The maximum average annual rate of soil erosion (whether from rainfall or wind) that will permit a high level of crop productivity to be sustained economically and indefinitely. T factors of 1 through 5 are used, and these numbers represent the amount of soil loss in tons per acre per year that should be permitted on a given soil. The T factor is used in the soil-loss equations for rainfall and wind erosion.

Soil Map A map that shows the location and extent of the different soils in an area. The following kinds of soil maps are recognized: "detailed", "detailed reconnaissance", "reconnaissance", "generalized", and "schematic".

Soil Map, Detailed A soil map at a publication scale commonly of 2 inches = 1 mile on which soil types and soil phases are the main types of units delineated. The smallest unit size shown on such maps is about I 1/3 acres. Survey traverses are usually made at 1/4 mile or more frequent intervals. The unit boundaries on detailed soil maps should have been seen throughout their course and their placement on the map should be accurate to at least 50 to 100 feet. The maximum amount of unlike soil inclusions in mapped units is 15 percent.

Soil Map, Detailed Reconnaissance A reconnaissance soil map on which some areas or features are shown in greater detail than usual, or than others.

Soil Map, General (Generalized Soil Map) A soil map with a publication scale commonly of 1 inch - 1 mile on which soil associations and miscellaneous land types are the delineated units. The smallest unit size shown is about 3 1/2 acres. The maximum amount of unlike soil inclusions in such mapping units is 15 percent.

Soil Map, Reconnaissance Referring to a soil map of highly variable publication scale (1 inch = 1 mile to 1 inch = 8 miles) and quality. The most detailed units commonly shown are miscellaneous land types and soil associations or one or more phases of soil families. The smallest unit that can be shown on such a soil map at a scale of 1 inch = 1 mile is about 3 1/2 acres and at 1 inch = 4 miles about 280 acres. Map unit boundaries are plotted where they cross field-survey traverses. Traverses are at intervals varying from about 1/2 mile to several miles. Between these points of field observation most boundaries are sketched from the appearance of ground patterns on aerial photographs and the general appearance of the landscape.

Soil Mapping Unit A kind of soil, a combination of kinds of soil, or miscellaneous land type or types, that can be shown at the scale of mapping for the defined purposes and objectives of the survey. Soil mapping units are the basis for delineating a soil survey map. Mapping units normally contain inclusions of soils outside the limits of the taxonomic name, or names, used as the name for the mapping unit. Mapping units are generally designed only to reflect significant differences in use and management.

Soil Moisture (Soil Water) Water diffused in the soil immediately below the land surface (zone of aeration), from which water is discharged by transpiration in plants or by evaporation.

Soil Order The category at the highest level of generalization in the soil classification system. The properties selected to distinguish the orders are reflections of the degree of soil horizon development and the kinds of horizons present.

Soil Phase A subdivision of a soil taxonomic unit, usually a soil series or other unit of classification based on characteristics that affect the use and management of the soil but which do not vary sufficiently to differentiate it as a separate soil series. For example, a variation in a property or characteristic. such as degree of slope, degree of erosion, content of stones, texture of the surface, etc. Phases of soil series are the major components of the soil mapping units shown on detailed soil maps.

Soil Productivity The capacity of a soil to produce a specific crop as limited by its physical, chemical, and biological condition, the parent material, topographic position, climate of the site, and prior land use history.

Soil Profile The physical and chemical features of the soil imagined or seen in vertical section from its surface to the point at which the characteristics of the parent rock are not modified by surface weathering or soil processes such as leaching, oxidation, and accretion.

Soil Reaction The degree of acidity or alkalinity of the soil mass expressed in pH values or in words as follows: extremely acid, below 4.5; very strongly acid, 4.5-5.0; strongly acid, 5.1-5.5; medium acid, 5.6-6.0; slightly acid, 6.1-6.5; neutral, 6.6-7.3 (strictly 7.0); mildly alkaline, 7.4-8.0; strongly -alkaline, 8.1-9.0; very strongly alkaline, over 9.1.

Soil Series Soils which have similar soil profile characteristics (except for texture of the surface portion); and which are derived from similar parent materials are grouped together in a "soil series". A soil series is usually named for the locality where the typical soil was first recorded.

Soil Subgroup Soil great groups are subdivided into subgroups that 1. Show the central properties of the great group or.

2. Intergrade subgroups that show properties of more than one great group or.

3. Other subgroups for soils with atypical properties that are not characteristic of any great group.

Soil Survey A general term for the systematic examination of soils in the field and in laboratories; their description and classification; the mapping of kinds of soil; the interpretation of soils according to their adaptability for various crops, grasses, and trees; their behaviour under use or treatment for plant production or for other purposes; and their productivity under different management systems.

Soil Taxonomic Unit (Soil Taxon) A unit comprising all soils that fall within the defined limits of a class at any categorical level in a system of soil classification.

Soil Texture The relative proportions of the three individual particle-size classes (i.e., sand, silt, and clay) in a soil.

Soil Type 1. A subdivision of a soil series based on surface soil texture. At the present time a soil type is considered as a kind of soil phase and is not part of the soil classification system presently being used.

2. In Europe, a class roughly equivalent to a great soil group.

Soil Variant A kind of soil whose properties are believed to be sufficiently different from recognized soil series to justify a new series name but comprising such a limited geographic area that creation of a new series is not justified.

Solar backscatter ultraviolet radiometer (SBUV) Instrument that measures the vertical distribution and total ozone in the Earth's atmosphere. Data is used for the continuous monitoring of ozone distribution to estimate long-term trends. SBUV instruments are flown on NOAA polar-orbiting satellites.

Solar constant Aka total solar irradiance. The constant expressing the amount of solar radiation reaching the Earth from the sun, approximately 1370 watts per square meter It is not, in fact, truly constant and variations are detectable.

Solar cycle Eleven-year cycle of sunspots and solar flares that affects other solar indexes such as the solar output of ultraviolet radiation and the solar wind. The Earth's magnetic field, temperature, and ozone levels are affected by this cycle.

Solar radiation Energy received from the sun is solar radiation. The energy comes in many forms, such as visible light (that which we can see with our eyes). Other forms of radiation include radio waves, heat (infrared), ultraviolet waves, and x-rays. These forms are categorized within the electromagnetic spectrum.

Solar wind A continuous plasma stream expanding into interplanetary space from the sun's corona. The solar wind is present continuously in interplanetary space. After escaping from the gravitational field of the sun, this gas flows outward at a typical speed of 400 km per second to distances known to be beyond the orbit of Pluto. Besides affecting Earth's weather solar activity gives rise to a dramatic visual phenomena in our atmosphere. The streams of charged particles from the Sun interact the Earth's magnetic field like a generator to create current systems with electric potentials of as much as 100,000 volts. Charged electrons are energized by this process, sent along the magnetic field lines towards Earth's upper atmosphere, excite the gases present in the upper atmosphere and cause them to emit light which we call the auroras. The auroras are the northern (aurora borealis) and southern (aurora Australis) lights.

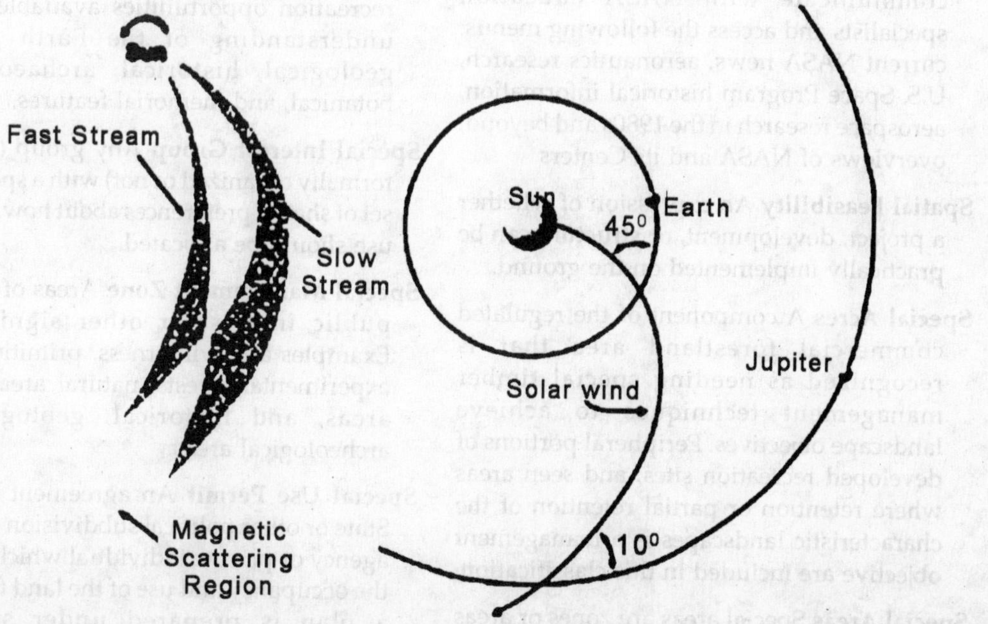

Fig. Solar wind

Solid Waste Debris and garbage generated by human activity.

Solitude An experience of individuals communing alone with their surroundings in a natural setting generally undisturbed by the sound or sight of technology; the state of being alone or remote from habitations; isolation; a lonely, unfrequented or secluded place.

Sounder A special kind of radiometer that measures changes in atmospheric temperature with height, as well as the content of various chemical species in the atmosphere at various levels. The High Resolution Infrared Radiation Sounder (HIRS), found on NOAA polar-orbiting satellites, is a passive instrument.

Space Environment Monitor (SEM) Instrument that measures the condition of the Earth's magnetic field and the solar activity and radiation around the spacecraft, and transmits these data to a central processing facility. NOAA polaForbiting and geostationary satellites both carry SEMs.

Space physics Scientific study of magnetic and electric phenomena that occur in outer space, in the upper atmosphere of the planets, and on the sun.

Space Shuttle NASA's manned, recoverable spacecraft designed to be used as a launch vehicle for Earth-orbiting experiments and as a short-term research platform.

Spacelab A manned laboratory module built by the European Space Agency (ESA) that accommodates dozens of experiments on each flight, mainly in the categories of materials science and life science.

Spacelink NASA electronic database for educators, with information stored on a computer at the Marshall Space Flight Center. Via computer educators communicate with NASA education specialists and access the following menus: current NASA news, aeronautics research, U.S. Space Program historical information, aerospace research in the 1980s and beyond, overviews of NASA and its Centers.

Spatial Feasibility An expression of whether a project, development, or structure can be practically implemented on the ground.

Special Acres A component of the regulated commercial forestland area that is recognized as needing special timber management techniques to achieve landscape objectives. Peripheral portions of developed recreation sites, and seen areas where retention or partial retention of the characteristic landscapes is a management objective are included in this classification.

Special Areas Special areas are zones or areas such as Wilderness, Research, Natural, or Scenic classified by Congress, the Secretary of Agriculture, Chief of the Forest Service, or Regional Forester. Areas within the National Forest System designated for their unique or special characteristics.

Special Cutting Logging activities in special areas, such as recreation areas and administrative sites, where other uses or other values override timber production values.

Special Exception (Exception) An exception (or special exception) is the granting of a permit for a use which is a deviation from standard zoning practice, but it is anticipated within the zoning ordinance and provisions for "exceptions" are made within the text. A private school, for instance, would not ordinarily be permitted in an area zoned exclusively residential; yet, since this might be a perfectly desirable and acceptable place for a school, allowance for the exception, subject to close scrutiny by the zoners, will be pre-established on the books.

Special Interest Area Areas managed to make recreation opportunities available for the understanding of the Earth and its geological, historical, archaeological, botanical, and memorial features.

Special Interest Group Any group (whether formally organized or not) with a specialized set of shared preferences about how resource use should be allocated.

Special Management Zone Areas of unusual public interest or other significance. Examples are: wilderness, primitive areas, experimental forests, natural areas, scenic areas, and historical, geological, or archeological areas.

Special Use Permit An agreement with the State or other political subdivision or public agency or private individual which permits the occupancy and use of the land for which a plan is prepared under specified conditions. Most permits require the payment of a fee.

Special Use Road A road constructed and used across National Forest lands under permission from the Forest Service.

Species 1. A unit of classification of plants and animals, consisting of the largest and most inclusive array of sexually reproducing and cross-fertilizing individuals which share a common gene pool.

2. The smallest natural population regarded as sufficiently different from all other populations to deserve a name, and assumed or proved to remain different despite interbreeding with related species.

3. Any native taxon of the plant or animal kingdom, including subspecies, distinct population segments, or designated evolutionarily significant units. Distinct population segments and evolutionarily

significant units are consistent with regulations developed by the Departments of the Interior and Commerce to implement the Endangered Species Act.. Any member of the currently accepted and scientifically defined plant or animal kingdoms of organisms.

Species Composition The kinds and numbers of species occupying a specified area. The list, count of species, (and often proportions) of (tree) species occurring together in the same stand.

Species Diversity (Biologic Diversity) The number of different species occurring in some location or under some condition, e.g., water pollution. Of the total number of species in a biotic community only a few are usually abundant while most are relatively uncommon. Because the large number of uncommon, relatively unimportant species largely determine the amount of "species diversity", this property is often expressed as a species diversity index which is calculated so as to better reflect the importance of those few species whose numbers, biomass, productivity, etc., so greatly dominate these attributes of the entire biological community. Species diversity tends to be low in physically controlled ecosystems (i.e., subjected to strong physiochemical limiting factors) and high in biologically controlled ecosystems. Diversity tends to be high in older communities and low in newly established ones. Diversity is directly correlated with ecological stability, but it is not certain to what extent this relationship is a cause-and-effect one.

Species Diversity Index (Diversity Index) The ratio between the total number of species in a biotic community and some rating of the relative importance (numbers, biomass, productivity, etc.) of individual species. Often the Shannon-Weiner index is used.

Species Richness The number of species (usually by groups such as birds or mammals) in any given area.

Species Viability A species consisting of self-sustaining and interacting populations that are well distributed through the species' range. Self-sustaining populations are those that are sufficiently abundant and have sufficient genetic diversity to display the array of life history strategies and forms to provide high likelihood for their long-term persistence and adaptability over time.

Species, Edge Species that occur primarily, most abundantly or most of the time in the transition zones between two or more biotic communities, i.e., in the ecotone.

Species, Endangered An animal or plant the prospects of which for survival and reproduction are in immediate jeopardy. Its peril may result from one or many causes — loss of habitat or change in habitat, overexploitation, predation, competition, disease or even unknown reasons. An endangered species must have help or extinction may follow. Any species which is in danger of extinction throughout all or a significant portion of its range- other than members of the class Insecta. Animals are declared "endangered" because their continued existence is threatened by one or more conditions. If any of the following conditions are true, the species (or subspecies) under consideration is declared "endangered": (a) the mortality rate consistently exceeds the birth rate; (b) it is incapable of adapting to environmental change; (c) its habitat is threaten by destruction or serious disturbance; (d) its survival is threatened by the unwanted introduction of other species through predation, competition, or disease; or (e) environmental pollution threatens its survival. Rare and endangered species are listed in the "Red Data Book" published by the International. (4) Those species in danger of extinction throughout all or a significant portion of their ranges. Species or subspecies from very limited areas, e.g., the type localities only, or from restricted fragile habitats usually are considered "endangered.".

Species, Endemic A species the natural occurrence of which is confined to a certain region and the distribution of which is relatively limited.

Species, Exotic Any species of wildlife not native to the country.

Species, Feral Non-native species or their progeny, which were once domesticated but have since escaped from captivity and are now living as wild animals, e.g., such as wild horses, burros, hogs, cats and dogs. Escaped plants are called "exotics.".

Species, Game 1. Wild animals, usually mammals or birds, hunted for sport or food and subject to legal regulations.

2. Any species of wildlife for which seasons and bag limits have been prescribed, and which are normally restricted to possession by sports-persons under state laws and regulations.

Species, Indigenous Any species of wildlife native to a given land or water area. By agreement, indigenous species may include introduced or exotic species that have established a niche in ecology of an area over a long period and are compatible with certain land management objectives. Hungarian and chukar partridges and ring-necked pheasants are examples.

Species, Introduced (Exotic Species) 1. Any species of wildlife not native to the country.

2. Any species that is not native in the area where it occurs.

Species, Key A species which plays an important ecological role in determining the overall structure and dynamic relationships within a biotic community. An evolutionary, component species of a biotic community whose presence is essential to the integrity and stability of a particular ecosystem. Key species may be unimportant as energy transformers in a biotic community (i.e., they may be neither very abundant nor consume large portions of the biotic productivity of a community). However, slight variations in key species' abundance results in large changes in the abundance of other species and/or in biotic community relationships and structure.

Species, Keystone A species which plays an important ecological role in determining the overall structure and dynamic relationships within a biotic community. An evolutionary, component species of a biotic community the presence of which is essential to the integrity and stability of a particular ecosystem. Keystone species may be unimportant as energy transformers in a biotic community (i.e., they may be neither very abundant nor consume large portions of the biotic productivity of a community). However slight variations in keystone species' abundance results in large changes in the abundance of other species and/or in biotic community relationships and structure.

Species, Local A species with a relatively small range but which is sufficiently common not to be called rare.

Species, Native Animals or plants which originated in the area in which they are found, (i.e., were not introduced and naturally occur in that area).

Species, Naturalized A (non-native) species established as if it were a native species, e.g., the annual grasses of California's Central Valley and foothills.

Species, Non-game A species not usually considered worthy of pursuit by sports-persons either for sport or food.

Species, Peripheral (Boundary Species) A species, or subspecies, the occurrence of which is at the edge of its natural range, and which is rare or endangered within the country, although not in its range as a whole. Special attention is necessary to ensure retention in the nation's fauna. A species which reaches the limit of its natural range a short distance into a country or some other boundary of significance to making planning area decisions.

Species, Rare One that, although not presently threatened with extinction, is in such small numbers throughout its range and widely separated in small sub-populations, so that inbreeding between sub-populations is seriously reduced, competition unlikely classes of rare species. 1. Occurring as a very few individuals or small groups at widely scattered localities over a large geographic area of what appears to be suitable habitat.

2. Found in very small numbers widely dispersed in each community where they grow, but which occur in many suitable areas over their geographic range.

3. Having a range restricted to so few localities that they are considered rare even though they occur in large numbers at each locality. Rare is the condition if: (a) it is confined to a relatively small and specialized habitat, and it is incapable of adapting to different environmental conditions; (b) although found in other parts of the world, it is nowhere abundant; (c) it is so limited that any appreciable reduction in range, numbers, or habitat would cause it to become endangered; or (d) if current management and protection programs were diminished in any degree, it would become endangered.

Species, Recently Extinct or Possibly Extinct Those species no longer known to exist after repeated search of the type localities and other known or likely places. Some species may be extinct in the wild but are being preserved by cultivation in gardens or as domesticated animals. Rare and endangered species are listed in the "Red Data Book" published by the International Union for Conservation of Nature and Natural Resources. Rare is the condition if: 1. It is confined to a relatively small and specialized habitat, and it is incapable of adapting to different environmental conditions;.

2. Although found in other parts of the world, it is nowhere abundant;.

3. It is so limited that any appreciable reduction in range, numbers, or habitat would cause it to become endangered; or.

4. If current management and protection programs were diminished in any degree, it would become endangered.

Species, Representative Representative species are those species which are representative, in terms of their biological requirements, of a balanced, indigenous community of wildlife in the study area of interest.

Species, Resident Species common to an area without distinction as to being native (i.e., native organism or endemic organism) or introduced (i.e., exotic species, naturalized species or feral species).

Species, Status Undetermined A species, or subspecies, that has been suggested as possibly being endangered but about which there is not enough information to determine its true status.

Species, Threatened Any species which is likely to become an endangered species within the foreseeable future throughout all or a significant portion of its range. This includes species categorized as rare, very rare, or depleted.

Species, Unique Species which are not endangered, but have considerable scientific, local, or national interest.

Species-of-Concern Species for which the Responsible Official determines that management actions may be necessary to prevent the species being listed under the Endangered Species Act.

Species-of-Interest Species for which the Responsible Official (US Forest Service) determines that management actions may be necessary or desirable to achieve ecological or other multiple use objectives.

Specific Conductance A measure of the ability of a water to conduct an electrical current; measured in micromhos per centimeter at 25 degrees centigrade.

Specific Source An artificial conduit or other conveyance where pollutants are discharged (from factories, sewage treatment plants, etc.) into a water body or aquifer.

Spectral band A finite segment of wavelengths in the electromagnetic spectrum.

Spectrum 1. The series of coloured bands diffracted and arranged in the order of their respective wave lengths by the passage of white light through a prism or other diffracting medium and shading continuously from red (produced by the longest visible wave) to violet (produced by the shortest visible wave).

2. Any of various arrangements of coloured ands or lines, together with invisible components at both ends of the spectrum, similarly formed by light from incandescent gases or other sources of radiant energy, which can be studied by a spectrograph.

3. In radio, the range of wave lengths of radio waves, from 3 centimeters to 30,000 meters, or of frequencies of radio waves, from 10 to 10,000,000 kilocycles. Also radio spectrum.

4. The entire range of radiant energies.

Spodosols Soils with alluvial accumulations of amorphous materials (organic matter and compounds of aluminum and usually iron) in subsurface horizons. These soils are formed in acid, mainly coarse-textured materials in humid and mostly cool or temperate climates. The particle-size class is mostly sandy, sandy- skeletal, coarse-loamy, loamy-skeletal, or coarse-silty.

Spoil Bank (Spoil Heap, Spoil Pile) A mound of mine refuse created by the deposit of overburden or non- coal material prior to backfilling the coal seam. Synonyms include cast overburden, tip, culm, gob, refuse pile, slate dump, stack, and heap. Includes acid spoil with pH below 4.0. Also refers to spoil having amounts of minerals such as aluminum, manganese, and iron that adversely affect plant growth.

Spoil The overburden or non-coal material removed in gaining access to the coal or mineral material in surface mining.

Spoil, Acid Usually spoil containing sufficient pyrite so that weathering produces acid water and the pH of the soil determined by standard methods of soils analysis is between 4.0 and 6.9.

Spot Fire Fire set outside the perimeter of the main fire by flying sparks or embers.

Spot Grazing Repeated grazing of small areas while adjacent areas are lightly grazed or unused.

Spraying Application of a liquid formulation (e.g., fertilizer, herbicide, insecticide) by air or ground delivery systems.

Spring A place where water flows from bedrock or soil upon the land surface or into a body of water.

Sprout A tree that grows from the stump or sucker root of a parent tree: it is not of seed origin. Basswood is frequently of sprout origin.

Stage (Complete) The stage of a stream or lake is the height of the water surface above an established datum plane.

Stand 1. An aggregation of trees or other growth occupying a specific area and sufficiently uniform in composition (species), age arrangement, and condition as to be distinguishable from the forest or other growth on adjoining areas (a minimum of 1 acre of forest land that is at least 10 percent stocked by forest trees of any size). A forest stand is said to be " pure" if 80 percent or more of the trees present are of the same species. If less than 80 percent of all trees present are of the same species, the stand is said to be " mixed.".

2. A plant and animal community.

Stand Density The quantity of trees per unit area. Density usually is evaluated in terms of basal area or percent-crown cover.

Standard A statement or a demonstration representing conditions of a job done properly. Standards show *how well* something should be done, rather than what should be done. Also criteria for desirable or tolerable conditions. A "standard" may not be fair, equitable, or based on sound scientific knowledge, (e.g., of the functional cause-effect relationship between specific pollutant concentrations and the resulting levels of damage) for it may have been established somewhat arbitrarily on the basis of best available technical data tempered by a factor for safety.

Standard Acres The regulated commercial forest land area on which crops of industrial wood can be grown and harvested with adequate protection of the forest resources under the usual provisions of the timber sale contract.

Standard Metropolitan Statistical Area A county or group of contiguous counties that contains at least one city of 50,000 inhabitants or more (or "twin cities" with a combined population of at least 50,000); also, contiguous counties essentially metropolitan in character and socially and economically integrated with the central city.

Standard of Living An organic whole, touching every phase of the life of the individual, the family, or the group. They concern the kind and quantity of food consumed, the sort of house lived in, clothing worn, etc., as well as the various cultural elements of existence, such as education, recreation, participation in church and civic organizations, and numerous other activities.

Standing Crop The total number or the total weight of one or more species of animals and/or plants in an area at a given time. Biomass relates best to weight per unit area; "standing crop" relates to the summation of all those weights per unit area over some spatial extent of concern. Biomass may mean phytomass (plants only).

Start tone Five seconds of 300 Hz black to white square wave modulation of the WEFAX subcarrier signaling the start of a frame transmission (the beginning of a direct readout image).

State Economic Area Single counties or groups of counties that have similar economic and social characteristics. The boundaries of these areas have been drawn in such a way that each state is subdivided into relatively few parts, with each part having certain significant characteristics that distinguish it from adjoining areas.

Station Point The location of the viewer looking upon an objective particularly with regard to being below, at the same level as, or above the visual objective.

Statistic A summary value calculated from a sample of observations, usually but not necessarily as an estimator of some population parameter; a function of sample values.

Steady-State Conditions The desired future environmental conditions to be created and maintained by using management practices as described in the management area prescriptions.

Steno- A prefix denoting an organism which can only tolerate a narrow range of changes of one or more of its environmental conditions. For example, stenothermal refers to organisms that cannot tolerate much change in temperature. Eury-organisms can tolerate large changes.

Stewardship A synonym for management, but it more specifically refers to resource management, and implies an emphasis on the careful and conscientious use and conservation of resources and ecosystems in a sustainable manner.

Stick, Biltmore A rule graduated in such a way that the diameter of standing tree may be estimated when the stick is held tangent to the surface and at right angles to the main axis of the tree, and at a distance from the eye for which the slick is graduated.

Stick, Scale A graduated stick for measuring the diameters and contents of logs; both measures are stamped on the stick.

Stochastic Process Any process the development of which in time is apparently governed by chance or probabilistic considerations.

Stocking Level In a forest, a more or less subjective indication of the number of existing trees (stem density) in a stand as compared to the desirable number for best growth and management, e.g., maximum productivity of wood. Expressions: well-stocked, overstocked, or partially stocked.

Stocking Plan The number and kind of livestock assigned to one or more given management areas or management units for a specified period.

Stocking Rate (Grazing Rate) The actual number of animals, expressed in either animal units or animal unit months, on a specific area at a specific time.

Stocking, Proper Placing a number of animals on a given area that will result in proper use at the end of the planned grazing period. Continued proper stocking leads to proper grazing.

Stomata Minute openings on the surfaces of leaves and stems through which gases (e.g., oxygen, carbon dioxide, water vapour) and some dissolved minerals pass into and out of plants.

Stop tone Five seconds of 450 Hz black to white square wave modulation of the WEFAKsubcarrier signaling the stop of a frame transmission (end of a direct readout image).

Storage Coefficient The volume of water released from storage in a vertical column of 1.0 square foot, when the water table or piezometric surface declines 1 foot.

Stored Function A mathematical function that is predefined by the Data Base Administrator and included as part of the schema. It is invoked as part of an Action Clause in order to perform some calculation.

Storie Index A numerical expression of the degree to which a particular soil presents conditions favourable for plant growth and crop production under favourable climate and adequate moisture conditions. Four general factors are considered: 1. The character of the soil profile;.

2. Soil texture;.

3. Slope of the land and;.

4. Other modifying factors such as drainage, salinity, alkalinity, soil acidity, erosion conditions, fertility level and microrelief. Each of these four factors is evaluated on the basis of 100 percent for the most favourable or ideal conditions, with limiting maximum and minimum ratings ascribed to conditions that are less favourable for plant growth.

Strategy A carefully considered plan or method, more encompassing and larger scale than tactics, for achieving an objective.

Stratification (Land Stratification) Grouping land units into similar units to reduce their variance and increase economy of sampling effort and descriptions. A planning unit can be split into homogeneous response units (without names or classes) about which useful generalizations can be made without having to make site specific studies.

Stratification, Chemical 1. A condition found in temperate lakes of the second order during the summer and winter stagnation periods in which certain horizontal strata become different chemically from adjacent ones, often with abrupt transitions.

2. Layering of water in a lake because of density differences owing to the varying or differential concentrations of dissolved substances with depth.

Stratification, Thermal The layering of water masses owing to different densities in response to temperature. The condition of a body of water in which the successive horizontal layers have different temperatures, each layer. more or less sharply differentiated from the adjacent ones, the warmest (or the coldest) at the top.

Stratified Seed Seed that has been stoned in a cool, moist condition before use. This storage practice hastens the germination of some species.

Stratosphere Region of the atmosphere between the troposphere and mososphere, having a lower boundary of approximately 8 km at the poles to 15 km at the equator and an upper boundary of approximately 50 km. Depending upon latitude and season, the temperature in the lower stratosphere can increase, be isothermal, or even decrease with altitude, but the temperature in the upper stratosphere generally increases with height due to absorption of solar radiation by ozone.

Stream A body of water flowing in a natural channel as distinct from a canal.

Stream Flow The discharge of water from a watershed that occurs in a natural stream channel.

Stream Load The quantity of solid and dissolved material carried by a stream.

Stream, Intermittent A stream which flows only part of the year.

Stream, Interrupted A stream in which part of the flow is underground.

Stream, Perennial A stream which flows throughout the year with waters covering the substratum.

Streamcourses Any stream bed carrying intermittent or permanent flows of sufficient volume, frequency, or velocity to develop a clearly defined channel.

Stream-gauging Station A gauging station where a record of discharge of a stream is obtained. Within the Geological Survey this term is used only for those gauging stations where a continuous record of discharge is obtained.

Streamside Management Zone (SMZ) An area of natural timber or vegetation protected and maintained on each side of a stream or drainage to provide habitat diversity and wildlife corridors, and to protect water quality.

Stress 1. The conditions resulting from any environmental change that disturbs the normal functioning of an animal to such an extent that its chances for survival are reduced.

2. The result or consequent state of a physical, chemical, or social stimulus on an organism or system;.

3. a state of strain, resulting from stress; a stimulus, but medical ecology uses "stressor.".

4. The condition of an animal in which large parts of the body deviate from their normal

resting state, either because of their activity or because of an injury.

5. The total energy with which water is held in the soil.

Stress, Thermal The heat tolerances affecting organisms.

Strip Cutting A clear cutting system variation in which the logging operation removes all merchantable timber from areas that run through a stand and are usually of a width equal to one or two times the general stand height.

Strip Pit A coal or other mine, usually open-pit, worked by stripping away the surface soil and rock above a mineral deposit.

Structural Diversity The degree of variety, vertically or horizontally, perhaps volumetrically, in the herb, shrub, and tree layers of vegetation of an area.

Structural Range Improvement Any type of range improvement that is made by people (fences, corrals, etc.).

Stumpage Price The price a logger is willing to pay for wood as it is in the woodland or "on the stump."

Stumpage Uncut trees standing in the forest. Sometimes used to mean the commercial value of standing trees.

Stumpage Value The value of timber as it stands uncut, on the stump, in terms of an amount of money per cubic unit. The value is usually small since likely local costs of overhead, cutting, and transportation must be added before the tree reaches market and is bought by someone.

Subcarrier The 2400 Hz audio tone transmitted by APT and WEFAX spacecraft. Amplitude modulation of this tone is used to convey video information.

Subclimax A stage in succession prior to the climax community in which further development is inhibited by the influence of some factors other than climatic factors.

Subculture (Group Culture) A group with different folkways and mores developed within a society. Groups with common play, age, vocational, sexual, habitational, or other interests evolve behaviour patterns that differ from those of other groups and from society's conventions.

Subdivision (Land Subdivision) The process (or area itself) of dividing a given area of undeveloped land into individual home sites and/or blocks of lots with streets or roads and open spaces. (After Abrams 1971).

Submarginal Land Land which is economically incapable of sustaining indefinitely a certain use or ownership status.

Suboptimization The selection of the best alternative course of action which pertains to a subproblem, that is, to only part of the overall problem or objective. Suboptimization is usually necessary because alternatives at all the various levels of decision making cannot, as a practical matter, be analyzed simultaneously before decisions are made at any level.

Subsatellite point Point where a straight line drawn from a satellite to the center of the Earth intersects the Earth's surface.

Subsidence A downward movement of the ground surface caused by solution and collapse of underlying soluble deposits, rearrangement of particles upon removal of underground mineral deposits, reduction of fluid pressures within an aquifer or petroleum reservoir, or collapse of land over mined lands.

Subsoil The B soil horizon of soils with distinct soil profiles. In soils with weak profile development, the subsoil can be defined as that below the soil in which roots normally grow (i.e., below the plowed soil or its equivalent of surface soil). It cannot be defined accurately; a carry-over from early days when "soil" was conceived only as the plowed soil and that under it as the "subsoil".

Substratum The surface on or in which organisms live. Substrate is chemical term referring to a compound or molecule upon which a reaction takes place. The layer on which organisms grow, often used synonymously with surface of ground. The substance, usually a protein, attached by an enzyme; often but improperly used variant of substratum.

Subsystem 1. A subunit of either the physical climate system (e.g., ocean dynamics) or the biogeochemical cycles (e.g., terrestrial ecosystems).

2. A subunit of a spacecraft, e.g., the telemetry subsystem, the power subsystem, the sensor subsystem, etc.

Succession (Ecological Succession, Ecosystem Development) An orderly process of biotic community change or development that involves changes in species, structure, and community processes with time. The gradual replacement of one plant community by another. It is reasonably directional and, therefore, largely predictable. It results from modification of the physical environment by the community. Succession is community-controlled even though the physical environment determines the pattern, the rate of change, and often sets limits as to how far development can go. It culminates in a stabilized ecosystem in which maximum biomass and symbiotic function between organisms are maintained per unit of available energy flow. The whole sequence of communities that replaces one another in a given area is called the sere; the relatively transitory communities are variously called "seral stages" or "developmental stages." "Pioneer stages" occur first; the terminal stabilized stage is the climax. Species replacement in the sere occurs because populations tend to modify the environment, making conditions favourable for other populations until equilibrium between biotic and abiotic is achieved. A "sere" whose first stage is open water is termed a "hydrosere", one whose first stage is dry ground, a "xerosere" Succession is "primary" on sites that have not previously borne vegetation, "secondary" after the whole or part of the original vegetation has been supplanted, "allogenic" when the causes of succession are external to and independent of the community (e.g., accretion of soil by wind or water, or a change of climate) and "autogenic" when the developing vegetation is itself the cause.

Successional Stages. The different structural and compositional phases of vegetation development of forests and grasslands that occur over time and include the major developmental or seral stages that occur within a particular environment.

Succession, Primary Successions beginning from newly formed soils or upon surfaces exposed for the first time (as by land slides or lava) which have as a consequence never borne vegetation or animals.

Succession, Secondary Any succession caused by a human agency following the destruction of part or all of the vegetation in an area.

Sugar Bush A stand mostly of sugar maple trees which is currently used for gathering sap for producing maple syrup.

Suitability (Capability) Rating of use or productivity potentials based on the present state of the resource. Evaluation based on the resource's inherent, natural or intrinsic ability to provide for use and includes that existing ability that is the result of past alterations or current management practices. Another evaluation procedure rates the potential ability of a resource to produce goods or services on the basis of the maximum possible outputs for a given type and level of future, alternative site, or resource management inputs. G. Angus Hills used the terms use capability to refer to evaluations based on a resource's inherent or present condition abilities, use suitability for evaluations based on potential management inputs, and use feasibility to refer to usability potential ratings based on an evaluation of off site factors, e.g., such as accessibility, present and forecasted socioeconomic conditions, technological developments, etc. "Capability" has precedence and should be used when referring to evaluations for usability based on the present state of the resource and "suitability" should only be used for evaluations based on assumptions about potential usability or productivity if specified management alterations were to be made. Modifiers should accompany both terms.

The appropriateness of applying certain resource management practices to a particular area of land, as determined by an analysis of the social, economic, and environmental consequences and the alternative uses foregone. A unit of land may be suitable for a variety of individual or combined management practices.

Suitability, Intrinsic Relative display of locations within a planning area having the maximum coincidence of favourable natural environmental factors relevant to the location of a prospective land use and the fewest constraints on that use.

Suitable Forest Land Land that is managed for timber production.

Suitable Range Area which is accessible to livestock, is producing or has the potential of producing an adequate amount of forage that can be grazed by livestock on a sustained yield basis under an intensive management system without causing damage to the range, wildlife habitat, timber regeneration, soil, watershed or other resource values.

Suitable Timber Lands Forest lands to be managed for timber production.

Summer Annual Grasses Grasses that must be replanted each spring. These plantings provide summer feeding areas for many kinds of wildlife, especially young turkeys and quail which utilize green forage and insects. Plants that mature and produce seed in late surnmer- to early fall such as millets and sorghums are also used by seed-eating birds.

Summer Perennial Grasses Grasses that do not need to be replanted each spring. These plantings also provide green forage, seeds, and insects for many kinds of wildlife.

Sun The closest star to Earth (149,599,000 km away on average). The sun dwarfs the other bodies in the solar system, representing approximately 99.86 percent of all the mass in the solar system. One hundred and nine Earths would be required to fit across the Sun's disk, its interior could hold over 1.3 million Earths.

The source of the Sun's energy is the nuclear reactions that occur in its core. There, at temperatures of 15 million degrees Celsius (27 million degrees Fahrenheit) hydrogen atom nuclei, called protons, are fused and become helium atom nuclei. The energy produced through fusion at the core moves outward, first in the form of electromagnetic radiation called photons. Next, energy moves upward in photon heated solar gas-this type of energy transport is called convection. Convective motions within the solar interior generate magnetic fields that emerge at the surface as sunspots and loops of hot gas called prominences. Most solar energy finally escapes from a thin layer of the Sun's atmosphere called the photosphere-the part of the Sun observable to the naked eye.

The sun appears to have been active for 4.6 billion years and has enough fuel for another 5 billion years or so. At the end of its life, the Sun will start to fuse helium into heavier elements and begin to swell up, ultimately growing so large that it will swallow Earth. After a billion years as a "red giant" it will suddenly collapse into a "white dwarf." It may take a trillion years to cool off completely.

Sun synchronous Describes the orbit of a satellite that provides consistent lighting of the Earth-scan view. The satellite passes the equator and each latitude at the same time each day. For example, a satellite's sun-synchronous orbit might cross the equator twelve times a day each time at 3:00 p.m. local time. The orbital plane of a sun-synchronous orbit must also precess (rotate) approximately one degree each day, eastward, to keep pace with the Earth's revolution around the sun.

Supplement 1. Something that makes an addition; something that completes, adds a finishing touch, or brings closer to completion or a desired state.

2. A "complement" is a supplement that results in mutual benefits to a plan and its addition. Supplements do not gain anything unto themselves by their addition to a plan.

Supply Schedule (Curve) A schedule of amounts of an output that producers are willing to provide at a range of prices, at a given point in time and condition of sale.

Supply Schedule (Supply) The amount of a good or service that will be offered for sale over a given range of prices at a particular point in time.

Supply The amount of an output that producers are willing to provide at the specified price, time period, and condition of sale.

Suppressed The condition of a tree characterized by low growth rate and low vigor due to competition with overtopping trees.

Suppression Activities required in extinguishing or confining a fire, beginning with its discovery.

Surface Mining, Block Cut Method A method of surface mining in which overburden is removed and placed around the periphery of a box- shaped cut. After coal is removed, the spoil is pushed back into the cut and the surface is blended into the topography.

Surface Mining, Extended Bench Method A method of surface mining which employs a large capacity walking dragline in deep overburden operating from a machine-supporting bench formed by filling the "V" between the bank and the spoil. This V-section is formed from material that falls from the bank or is placed by the dragline and it must be rehandled. Rehandling averages approximately 10-25% of the solid bank, depending on the height of the bench.

Surface Mining, Lateral Movement Mining A method of surface mining in which coal is removed by stripping and augering with no material being placed on the downslope. Lateral movement reduces disturbed acreage by nearly two-thirds when compared with conventional surface mining because the overburden is hauled by truck laterally along the bench and then backfilled against the high- way.

Surface Mining, Modified Block Method This method of surface mining adapts the "Block cut" method to steeply sloped areas. The modi-fication essentially is backfilling with spoil from succeeding blocks rather than from the spoil-pro-ducing block.

Surface Mining, Open Pit Method A type of surface mining in which the overburden is removed from the product being mined and is dumped back after mining; or may specifically refer to an area from which the overburden has been removed and has not been filled.

Surface Mining, Strip Surface mining in which the over- burden is removed in narrow, consecutive bands, producing an uneven surface of ridges resembling a plowed field.

Surface Rights The ownership of or rights to the surface of the land only and the mineral rights have been separately reserved. Also, the right of a mineral owner to use so much of the surface of the land as may be reasonably necessary for the conduct of mining operations.

Surface Soil (Topsoil) The uppermost part of the soil ordinarily moved in tillage (or its equivalent in uncultivated soils) ranging in depth from 5 to 8 inches. It is frequently designated as the "plow layer," the Ap layer, or the Ap soil horizon.

Surface Storage The sum of detention storage and channel storage, representing, at any given moment, the total water enroute to an outlet from an area or watershed.

Surface Water Telemetry The surface water telemetry component identifies the type of system or equipment being used to transmit surface water information (primarily stage data) from the data collecting site to a central receiving site.

Surface Water Water on the surface of the Earth such as that in a river or lake.

Survey mode Refers to observational emphasis upon frequent global coverage, usually with restricted spatial and spectral resolution, aimed at developing a consistent, long-term data product for later interpretation.

Suspended Load The smaller suspended sediment particles that are lifted far from the bottom of a water body for long periods by running water and are distributed through the whole body of the current.

Suspended Sediment Sediment that remains suspended in the water for a long period of time without coming in contact with the bottom.

Suspended Solids Colloidal and particulate matter such as clay, sand, and finely divided organic material, that cause a cloudy condition and a term synonymous with *suspended sediment*; however, it is generally used by sanitary engineers in connection with water treatment facilities, while suspended sediment is generally used by civil or hydraulic engineers in connection with sediment transport studies.

Sustainability Meeting needs of the present generation without compromising the ability of future generations to meet their needs. Sustainability is composed of desirable ecological, economic, and social conditions or trends interacting at varying spatial and temporal scales. Embodies the principles of multiple-use and sustained-yield, appropriate to scale, without impairment to the productivity of the land.

Sustained Yield (Sustained Production) 1.An ideal forest management objective at which point the volume of wood removed is equal to growth within the total forest.Sustained-yield management, therefore, implies continuous production planned to achieve at the earliest practical time a balance between increment to the mature forest and its cutting.

2. Achieving and maintaining in perpetuity a nearly equal annual or regular periodic output of the various renewable resources, without impairment of the productivity of the land. Whether "yield" should be of wood or products ... or benefits...or profit...remains debated.

Sustained Yield Management Continuous production of the forest crop so planned as to achieve, at the earliest practical time, a balance between growth and cutting.

Swamp An area saturated with water throughout much of the year, but with the surface of the soil usually not deeply submerged.

Sweep Tree defect resulting from a gradual curve in the main stem of the tree.

Synergism The combined action of two or more agents the result of which is greater than the sum of the actions of each of the agents acting alone.

Synoptic chart Chart showing meteorological conditions over a region at a given time; weather map.

Synoptic view The ability to see large areas at the same time.

Synthetic aperture radar (SAR) A high-resolution ground-mapping technique that effectively synthesizes a large receiving antenna by processing the phase of the reflected radar return. The along4rack resolution is obtained by timing the radar return (tim~gating) as for ordinary radar The crosstrack (azimuthal) resolution is obtained by processing the Doppler phase of the radar return. The cross4rack "dimension" of the antenna is a function of the length of time over which the Doppler phase is collected. See Doppler effect.

System An organized collection of parts that are related to each other, united by regulated relations in such a way that changes in one component can affect some, or all of the other components.

System Function A predefined operator that enables the user to obtain from a computer or computed value such as average, count, or standard deviation of selected values.

System Road A road under a single administration on which regular programs of maintenance are carried out. All traffic is subject to State traffic laws.

Systems Analysis 1. The study of how component parts of a system interact and contribute to that system.

2. The act, process, or profession of studying an activity (as a procedure, a business, or a physiological function) typically by mathematical means in order to define its goals or purposes and to discover operations and procedures for accomplishing them most efficiently.

Systems Approach A systematic consideration of problems wherein analysis is stressed as well as intuition. a) Implies seeking an optimal course of action, i.e., objectives, costs, effectiveness, and risks of alternative strategies. This latter refers to systems analysis. b) Utilizing a systems approach to problems that entails (i) compiling, condensing, and synthesizing a great amount of information concerning the components of the system, (ii) examining in detail the structure of the system, (iii) translating this knowledge of systems components, function, and structure into models of the system, and (iv) using the models to derive new insights about managing and utilizing the systems.

T

Taproot The main root of a tree which strikes downward with or without heavy branching until it either reaches an impenetrable layer or one so lacking in oxygen or moisture that further downward growth is impossible.

Target A clear and concise statement used to express planned results to be reached within a stated time period. Often one or more separate targets are used. These are system performance measures.

Taxonomy The science of classifying organisms; the arrangement of organisms into systematic groups such as species, genus, family, and order. It is the theoretical study of classification including its bases, principles, procedures, and rules.

Technological Assessment (Technological Forecasting) The attempt to estimate or predict the completely new technological developments which will occur in the future; and, to evaluate the extent to which such likely, future new abilities will be able to solve (or at least acceptably minimize) current land use problems, or problems which are expected to be associated with particular land uses in the future.

Technological Change The increase in society's ability to solve problems or increase efficiency by technical means with time. Availability of alternative or completely new technical methods for solving problems associated with land uses and the inevitable evolution and development of new ideas.

Technological Solution The application of the devices or techniques developed by science for practical purposes. Often these are completely new technical methods for solving current or potential problems for which no "technological solutions" presently exist. The distinction between "technical solutions" and "engineering solutions" is that the first refers just to the development of new technical means while the last applies to the application of any technical means (new or old) to solving potential problems.

Telemetry Automatic transmittal of data to a point remote from the sensing instrument. Also use of small radios to monitor animal location, movement, behaviour, and physiology.

Temperature Measure of the energy in a substance. The more heat energy in the substance, the higher the temperature. The Earth receives only one twobillionth of the energy the sun produces. Much of the energy that hits the Earth is reflected back into space. Most of the energy that isn't reflected is absorbed by the Earth's surface. As the surface warms, it also warms the air above it.

Temporary Facility (Road) A temporary use of the land for transportation purposes; usually constructed to serve a specific resource activity. Occupied land is returned to resource production upon termination of the activity requiring its use.

Temporary Permit (Temporary License) A document authorizing grazing a certain number of livestock on public lands during an emergency or for a certain period, terminable at the end of such period and with no guarantee of renewal in whole or in part.

Terabit A trillion (1012) bits.

Term Permit (Term License) A document that authorizes grazing on public lands for a stated number of years as contrasted with an annual or temporary permit.

Terminal Facilities Parts of the transportation network and interface between the transportation network and resources served or between different transportation modes, e.g., vehicle parking areas, boat ramps and docks, trail heads, log landings and dumps, and air fields and heliports.

Terrace A level plain bordering a stream, river, or lake.

Terrace Threshold Sensitivity Level Sloping ground cut into a succession of benches and steep inclines for purposes of cultivation or to control surface runoff and to minimize soil erosion.

Terrace, Asorptite A ridge type of terrace used primarily for moisture conservation.

Terrace, Bench A terrace approximately on the contour, having a steep or vertical drop to the slope below, and having a horizontal or gentle sloping part.

Terrace, Drainage A broad channel-type terrace used primarily to conduct water from an area at a low velocity. It is adapted to less absorptive soil and high rainfall.

Terrain A tract of country considered with regard to its natural features, configuration, etc.; especially as affecting its advantages, fitness for use.

Terrain Analysis The process of interpreting a geographical area to determine the effect of the natural and human-made features on a particular activity, actual or proposed. This interpretation includes the influence of weather and climate on those features.

Terrain Factor Terrain factors are attributes or land forms, drainage features, ground, vegetation, etc. and cultural features or human-made changes in the Earth's surface.

Terrain Studies Regional studies usually with an emphasis on relief and essential physical features, e.g., soil, vegetation, and drainage.

Territoriality The behaviour by which an organism claims an area and displays, threatens, and defends it against members of its own species.

Territory The area which an animal defends, usually during breeding season, against intruders of its own species. Territories are smaller than and are normally located within an animal's home range.

Tertiary Facility Tertiary transportation facilities are usually intended to provide access for a specific resource utilization activity, such as a timber sale or recreation site, although other minor uses may be served. Emphasis is given to resource service rather than travel efficiency. Tertiary facilities may often be developed and operated for short term or intermittent service.

Tertiary Wastewater Treatment (Tertiary Treatment, Tertiary Sewage Treatment) Wastewater treatment processes, beyond primary and secondary treatment, which remove or reduce the amount of nitrate and phosphate nutrients present in their effluents. Tertiary treatment may consist of extensions or modifications of secondary treatment processes, additional forms of chemical treatment, electrochemical processing, carbon filtration, and other more complex procedures. Tertiary treatment removes up to 99 percent of the suspended solids and biological oxygen demand, 95 percent of the phosphate and 75 percent of the nitrate from wastewater.

Thematic mapper (TM) A LANDSAT multispectral scanner designed to acquire data to categorize the Earth's surface. Particular emphasis was placed on agricultural applications and identification of land use. The scanner continuously scans the surface of the Earth, simultaneously acquiring data in seven spectral channels. Overlaying two or more bands produces a false colour image. The ground resolution of the six visible and shortwave bands of the Thematic Mapper is 30 meters, and the resolution of the thermal infrared band is 120 meters. Thematic mappers have been flown on LANDSATs-4 and -5.

Thermal Cover, Structural For big game (animal) a stand of coniferous trees 40 feet tall or taller with average crown closure of 70 percent or more. In some cases, topography or vegetation less than specified may meet animal needs for thermal regulation. Functional: Situations, usually related to vegetation structure, used by animals to ameliorate effects of weather.

Thermal infrared Electromagnetic radiation with wavelengths between about 3 and 25 micrometers.

Thermal Pollution (Calefaction) Altering the amount of energy available in the environment by adding heat or cold. Most frequently this occurs in the form of the heat added by returning water which has been used for cooling purposes to a river, lake, estuary, bay or the ocean. However "thermal pollution" may also occur with the release of large volumes of the cold, deep water in reservoirs into warmer bodies of water.

Thermocline The layer of water in a lake between the epilimnion and hypo-limnion in which the temperature exhibits the greatest difference in Minimum level of exposure to a specific pollutant which results in a detectable effect. The important factor is to differentiate between visible and non-visible (physio-logical-internal) effects.

Thinning Cutting made in an immature crop or stand in order to accelerate the diameter increment (annual growth) of the residual trees but also, by suitabie selection, to improve the average form of the trees that remain, without (at least according to classical concepts) permanently breaking the canopy. Commonly expressed as "...to a basal area per acre of X amount for all trees over 5 inches in diameter.".

Thixotrophy Soil liquefaction.

Threatened and Endangered Species A species or subspecies of animals or plants whose prospects of survival and reproduction are in immediate jeopardy (endangered), or likely to become so within the foreseeable future (threatened).

Threatened Species Any species that is likely to become an endangered species within the foreseeable future throughout all or a significant portion of its range.

Threshold Level The minimum concentration or amount of a given substance or condition necessary to produce a measurable physiological or psychological effect. Also, the point at which a major change occurs, e.g., the change of water to ice at zero degrees centigrade.

Thunder The sound that results from lightning. Lightning bolts (static electricity) produce intense heat. This burst of heat makes the air around the bolt expand explosively producing the sound we hear as thunder.

Since light travels faster than sound, we see the lightning before we hear the thunder.

Thunderstorm Local storm resulting from warm humid air rising in an unstable environment. Air may start moving upward because of unequal surface heating, the lifting of warm air along a frontal zone, or diverging upper-level winds (these diverging winds draw air up beneath them). The scattered thunderstorms that develop in the summer are called airmass thunderstorms because they form in warm, maritime tropical air masses away from other weather fronts. More violent severe thunderstorms form in areas with a strong vertical wind shear that forces the updraft into the mature stage, the most intense stage of the thunderstorm. Severe thunderstorms can produce large hail, forceful winds, flash floods, and tornadoes.

Tiel Gradient The temperature difference within a layer 1 meter thick.

Tiering Incorporating information contained in an Environmental Impact Statement, such as the Forest Plan EIS, by reference in subsequent environmental documents.

Tilth The physical condition of a soil in respect to its fitness for the growth of a specified plant.

Timber Harvest The removal of trees for wood fibres utilization.

Timber Inventory A collection of information about a timber stand made by measuring tree and stand characteristics such as tree volume and grade and stand density.

Timber Management Compartment A small subdivision of forest area for purposes of orientation, administration, and silvicultural operations, and defined by permanent boundaries (either of natural features or artificially marked) which are not necessarily coincident with stand boundaries.

Timber Production 1. Purposefully growing, tending, harvesting, and regenerating of regulated crops of trees to be cut into logs, bolts, or other round sections for industrial or consumer use. The term does not include production of fuelwood or chip products.

2. The sustained long-term and periodic harvest of wood fibre from Forest System lands undertaken in support of social and economic objectives identified in one or more land and resource management plans. For purposes of this rule, the term timber production includes fuel wood.

Timber Resource System The role of a timber resource system is to grow and make available wood for the nation on a continuing basis. Thus, the system includes those activities necessary to 1. Protect, improve, grow, and harvest timber from forest land and.

2. Protect, process, and utilize wood and wood-related products. In addition to wood, the system produces other goods and services, either by design or incidentally.

Timber Sale Issuance of a contract for cutting and removing designated trees; the site and or work being done under such a contract.

Timber Sale Program Quantity The volume of timber planned for sale during the first decade of the planning horizon. It includes the allowable sale quantity (chargeable volume) and any additional material (non-chargeable volume) planned for sale. The timber sale program quantity is usually expressed as an annual average for the first decade.

Timber Site Index A measure of site productivity based on tree height growth over 50 years.

Timber Stand A community of trees sufficiently uniform in composition of species, size, age, etc. to be distinguished from adjacent communities. Maximum and minimum sizes are usually established to provide for other resource coordination or practicable timber management.

Timber Stand Improvement (TSI) All noncommercial intermediate cuttings (i.e., removing less desirable trees, vines and, occasionally large shrubs) and other practices to improve the composition, constitution, condition, stocking, and increment of a residual forest stand.

Timber Standing trees, usually of commercial size.

Time of Travel The rate of movement of water, or waterborne materials, through a defined reach of stream channel for steady or gradually varied flow conditions. Determined by dye tracing methods where dye is injected at some location on a stream and detected at other locations downstream (any time-distance measure).

Tolerable Soil Loss The amount of soil that can be lost over 100 years from accelerated erosion and not significantly or permanently lower soil productivity potential.

Tolerance 1.Resistance of an organism to high levels or excesses of a substance or the deficiency of an element or a condition in its environment e.g., elm trees are "resistant" to periodic flooding.

2. The capacity of a tree to develop and grow in the shade of and in competition with other trees. Trees able grow in full or partial shade are considered "tolerant." Trees requiring full sunlight for survival are considered "intolerant. ".

Tolerance Limit, Median The concentration of some toxic substance at which just 50% of test animals are able to survive for a specified period of exposure.

Tolerance, Ecological 1. The positive response of an organism to the excess or the deficiency of an element or a condition in its environment.

2. The capacity of an organism to endure or adapt to (usually temporary) unfavourable environmental factors.

3. An organism's capability to withstand a condition indefinitely and through all life stages. Short-term "tolerance" should be redefined in terms of stress.

Topography The configuration and relations of a surface including its relief, elevation and the position of its natural and human-made features. The physical or natural features of an object or entity and their structural relationships.

Topsoil Earthy material used as top-dressing for house lots, grounds for large buildings, gardens, road cuts, or similar areas. It has favourable characteristics for production of desired kinds of vegetation or can be made favourable. Also the surface plow layer of a soil. The original or present dark coloured upper soil that ranges from a mere fraction of an inch to two or three feet thick on different kinds of soil. The original "A" soil layer varying widely among different kinds of soil. Applied to soils in the field, the term has no precise meaning unless defined as to depth or productivity in relation to a specific kind of soil. Fertile soil material that is used as a top dressing. Distinction has been made among synthetic, weathered, and geologic topsoil. Synthetic topsoil can include sand and stone chips as well as fly ash, sawdust, or manure not usually a part of geological soil and rock. Weathered topsoil is the natural top-dressing material that has been subjected to all ecological processes through a long period.

Tornado A twisting, spinning funnel of low pressure air. The most unpredictable weather event, tornadoes are created during powerful thunderstorms. As a column of warm air rises, air rushes in at ground level and begins to spin. If the storm gathers energy a twisting, spinning funnel develops. Because of the funnel's cloud and rain composition and the dust, soil, and debris it draws up, the funnel appears blackish in colour. The most energetic storms result in the funnel touching the ground. In these tornadoes, the roaring winds in the funnel can reach 300 mph, the strongest winds on Earth. Funnels usually travel at 20 to 40 mph, moving toward the northeast. When tornadoes form over lakes or oceans they suck water into the funnel cloud and are called waterspouts.

Total Ozone Mapping Spectrometer (TOMS) Flown on NASAs Nimbus-7 satellite, its primary goal is to continue the high-resolution global mapping of total ozone on a daily basis. The Nimbus-7 launch in 1978 enabled TOMS to begin delivering data in 1979 and continue providing information until 1993. TOMS has mapped the total amount of ozone between the ground and the top of the atmosphere, provided the first maps of the ozone hole, and continues to monitor this phenomenon.

Because of its longevity, TOMS also has obtained information on the more subtle trends in ozone outside the ozone hole region. This results from development of a powerful new calibration technique that removes the instrument measurement drift that developed over the years. With this technique applied to the TOMS 14.5-year data record, a global ozone decrease of 2.69 percent per decade was detected.

To ensure that ozone data will be available through the next decade, NASA will continue the TOMS program using U.S. and foreign launches. In 1991, the former Soviet Union launched a Meteor-3 satellite carrying a TOMS instrument provided by NASA. A third TOMS will be launched onboard a NASA Earth probe satellite in 1994, and the Japanese Advanced Earth Observations Satellite (ADEOS) will carry a fourth TOMS when it launches in 1996.

Toxic A substance or the excessive amount of a substance that may cause death, stress, or impairment. Characteristics of things called poisons or toxins.

Toxicity Test A test that determines the potency of a toxic substance by measuring the intensity of a biological response.

Toxicity, Acute 1. Inhibition after an infinitely short exposure to a toxicant.

2. Toxicity which causes severe and rapid damage to the organism by the fastest acting mechanism of poisoning, fatal unless the organ-ism rapidly escapes the toxic environment.

Toxicity, Chronic Inhibition which may be lethal or sub-lethal after a long-term (relative to life cycle) exposure to a toxicant.

Toxicity, Lethal Concentration at which a toxicant causes death. 1. Concentration at which a toxicant first causes inhibition.

2. The lowest concentration (of a toxicant) that causes acute toxicity.

Toxicity, Unit Inhibition per log unit of toxicant at concentration greater than threshold toxicity.

Trace Elements Chemical elements appearing in minute quantities in natural systems or media; may occasionally be concentrated by specific organ-isms. Nutrients such as P, though in minute quantities, are not usually called trace elements.

Trace Elements, Essential - Toxic Effects A reduction in crop or ecosystem yield below that of the maximum yield plateau of a complete dose response curve and at concentrations greater than those required for maximum yield.

Trace Elements, Non-essential - Toxic Effects A reduction in yield below that of the zero dose level.

Tracking and Data Relay Satellite System (TDRSS) An orbiting communications satellite, developed by NASA, used to relay data from satellite sensors to ground stations and to track the satellites in orbit.

Trade winds Surface air from the horse latitudes that moves back toward the equator and is deflected by the Coriolis Force, causing the winds to blow from the Northeast in the Northern Hemisphere and from the Southeast in the Southern Hemisphere. These steady winds are called trade winds because they provided trade ships with an ocean route to the New World. See wind.

Trade-offs The combination of benefits and costs which are gained and lost in switching between alternative courses of action. "Trade-offs" include only those portions of benefits and costs that are *not* common to all alternative courses of action under consideration. The total estimated amount of resource program outputs (market and nonmarket) *foregone* by 1. Modifying an existing program or substituting one resource program for another or.

2. Due to switching among alternative courses of action. The economic value of this trade-off represents the Opportunity Cost.

Traffic Composition (Traffic Classification) The classification and grouping of the various sizes and weights of vehicles making up the traffic of a road such as passenger cars and trucks.

Traffic Service Levels Traffic service levels describe the significant traffic characteristics and operating conditions for a road.

Traffic Volume The number of vehicles traveling in both directions past a given point during a specified period of time. This is usually expressed as seasonal average daily traffic for roads that are open only part of the year, or annual aver age daily traffic for roads (highways) open year round.

Trail An 18-inch to 24-inch wide travel-way designed for pedestrian and/or horse riders. Also includes a travel-way up to 40 inches in width to accommodate off-road vehicles.

Trampling Damage to plants or soil brought about by foot movements or congestion of livestock and/or people.

Transitional Range Areas where big game concentrate during spring and/or fall. Transitional ranges are generally adjacent to winter range and may provide important security during the fall.

Transitory Range Rangeland created to increase forage production for livestock.

Transmission Pipeline A pipeline which carries gas or liquid from a producing field or central collection facility to a storage or consumption facility, usually over long distances.

Transmissivity (Water) The rate at which water of the prevailing kinematic viscosity is transmitted through a unit width under a unit hydraulic gradient.

Transpiration The loss of water in vapour form from a plant, mostly through the stomata and lenticels.

Transplant A tree which has been removed from its original seedbed and replanted one or more times in a nursery.

Transportation and Utility Corridor A parcel of land, without fixed limits or boundaries, which is being used as the location for one or more transportation or utility rights-of-way.

Transportation Network All existing and planned roads, trails, bridges, airfields, and other transport facilities wholly or partly within or adjacent to and serving a planning area.

Transportation Planning The process by which new transportation facilities (and improvements to the old facilities) are systematically designed, their adequacy tested, and their future construction programmed. Comprehensive transportation planning considers all modes (mass transit, automobile, air, rail, etc.) of transportation and treats them all as a single system.

Transportation Services Land Land used primarily for highways, roads, beltways, railroads, utility rights-of-way, airports, and other transportation facilities, together with necessary adjacent facilities such as approaches, underground and surface utilities, and other servicing structures, appurtenances, and measures.

Trasaction Costs The costs of bargaining, negotiating, and exchanging. These rise (time and difficulty) as more people with greater diversity become involved in decision making. Costs include those of identifying, locating, and contacting participants.

Travel Corridor An area adjacent to a road or trail which includes all of the land normally visible from the road or trail; a strip of land believed to be frequented by wild fauna.

Travel Influence Zone Areas comprising existing or anticipated significant public outdoor recreational occupancy, use, and enjoyment along existing and planned overland routes of travel; areas in and around existing or planned developed recreation sites. These are areas in which beauty of the landscape and other aesthetic values are key management considerations and are an important part of the out door environment. These areas offer significant opportunities for existing, planned, or anticipated recreational use and enjoyment by people visiting or traveling through the zone.

Tree A woody plant having a well-defined stem, more or less definitely formed crown and usually attaining a height of at least 10 feet.

Tree Cavities Hollow cavities in trees that provide resting or nesting places for wildlife.

Tree Farm A privately owned forest (woodland) dedicated to the production of timber crops.

Tree Injector Equipment specially designed to inject chemicals, usually phytocides, into the trunk of a tree.

Tree Shelter A plastic tube that can be wrapped around the stem of hardwood seedlings to increase survival and growth.

Trim Allowance Excess length of a log to allow for square trimming the lumber to an exact length.

Trophic Level In biotic communities organisms the food of which is obtained from the same part of the food chain. Green plants occupy the first trophic level, plant-eaters (i.e., herbivores) the second level, the carnivores (i.e., meat eaters) which eat the herbivores the third level, and secondary carnivores (i.e., carnivores which eat other carnivores) the fourth level.

Tropical Ocean-Global Atmosphere (TOGA) TOGA is a program jointly sponsored by the United Nations World Meteorological Organization (WMO); the International Council of Scientific Unions (ICSU); the United Nations Educational, Scientific, and Cultural Organization (UNESCO) Intergovernmental Oceanographic Commission (IOC); and the ICSU Scientific Committee on Oceanic Research (SCOR).

TOGA has four major objectives:.

To collect and catalog observations of the tropical atmosphere and ocean;.

To assess the evolution of the tropical atmosphere/ocean system in real time;.

To promote the development of short-term climate-prediction computer models for the tropics;.

To study the influence of the tropical atmosphere/ocean system on the climate at higher latitudes.

Tropical Rainfall Measuring Mission (TRMM) A joint NASA/NASDA mission planned for launch in 1997. The goal of TRMM is to obtain a minimum of 3 years of climatologically significant observations of rainfall in the tropics. Because rainfall is such a variable phenome non, adequate sampling is a difficult problem. By averaging the instantaneous rainfall rates for 30 days over a 50 by 50 grid, TRMM will obtain observations that meet climatological requirements. TRMM measurements, used together with cloud models, also will provide accurate estimates of vertical distributions of latent heating in the atmosphere.

The present uncertainty about the quantity and distribution of precipitation, especially in the tropics, prohibits definition of the mass and energy exchange between the tropical ocean and atmosphere. Since the tropical atmosphere and oceans are closely coupled, cloud radiation and rainfall are likely to have significant effects on ocean circulation and marine biomass.

TRMM data will play a significant role in global change studies, especially in developing an interdisciplinary understanding of atmospheric circulation, ocean-atmospheric coupling. and tropical biology TRMM data on tropical clouds, evaporation, and heat transfer will be used to understand the larger scale coupling of the atmosphere to oceans.

Tropical storm formation Tropical storms generally form in the eastern portion of tropical oceans and track westward. Hurricanes, typhoons, and willywuhes all start out as weak low pressure areas that form over warm tropical waters (e.g., surface water temperature of at least 80oF). Initially winds and cloud formations over the warm tropical waters are minimal. Both intensifv with time. Formation of tropical storms also requires a significant Coriolis effect to induce proper spin in the wind formation. As the storm begins to organize itself into a coherent pattern, it will experience increased activity and intensity.

When a storm develops a clearly recognizable pattern, it is referred to as a tropical depression. When wind speeds reach 35 knots (40.3 mph), it is called a tropical storm and is given a name. When wind speed equals or exceeds 74 mph, the storm is called a hurricane. In the western Pacific, a hurricane is referred to as a typhoon. In waters around Australia it is called a cyclone or willy-willy.

Hurricanes intensifv when moving over areas of increased water temperatures, and weaken over colder water surfaces. Upper atmosphere wind shear (different wind direction and speeds at different elevations) will frequently prevent or slow intensification of tropical storms by "spreading out" the storm horizontally and preventing the formation of strong updrafts of warm, humid air. Movement over a land-mass will weaken hurricane winds but will result in large-scale rain that can result in large-scale flooding. When encountering a strong frontal system (such as a polar front) the hurricane will curve and track along the leading edge of the front or become implanted in it.

Satellite infrared imagery can identifv surface water temperatures that will foster tropical storm development.

Tropics The area between 23.5 degrees north and south of the equator. This region has small daily and seasonal changes in temperature, but great seasonal changes in precipitation.

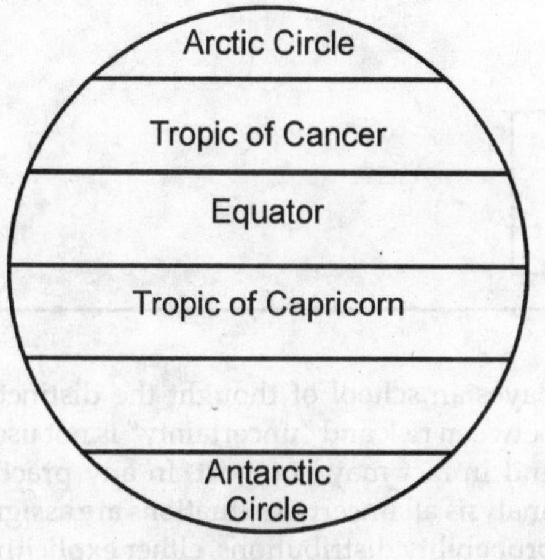

Fig. Tropics

Troposphere The lower atmosphere, to a height of 8-15 km above Earth, where temperature generally decreases with altitude, clouds form, precipitation occurs, and convection currents are active.

Tropospheric emission spectrometer A high-resolution infrared spectrometer for monitoring the minor components of the lower atmosphere.

Trough Elongated area of low atmospheric pressure, either at the surface or in the upper atmosphere.

True anomaly One of six Keplerian elements, it locates a satellite on an orbit. True anomaly is the true angular distance of a satellite (planet) from its perigee (perihelion) as seen from the center of the Earth (sun).

Trust Funds Funds collected and used by government for carrying out specific purposes and programs according to terms of a trust agreemenet or statute. Trust funds are administered by the government in a fiduciary capacity and are not available for the general purposes of the government.

Turbidity The capacity of materials suspended in water to cause light to scatter; it is measured in arbitrary Jackson Turbidity units (JTU). Highly turbid water is often called "muddy" although many kinds of suspended particles contribute to turbidity.

Turnout A widened space in-a road to allow vehicles to pass one another and which slopes away (downhill) from the road. Also, a drainage ditch which drains water away from roads.

Type Conversion The conversion of the dominant vegetation in an area from one species to another.

Typhoon Hurricanes in the Western Pacific Ocean.

U

U Ultisols Soils that are low in supply of bases and have subsurface horizons of alluvial clay accumulations. They are usually moist, but during the warm season of the year, some are dry part of the time. The particle-size class (texture) ranges mostly for sandy-loam to clayey.

Ultraviolet radiation The energy range just beyond the violet end of the visible spectrum. Although ultraviolet radiation constitutes only about 5 percent of the total energy emitted from the sun, it is the major energy source for the stratosphere and mesosphere, playing a dominant role in both energy balance and chemical composition.

Most ultraviolet radiation is blocked by Earth's atmosphere, but some solar ultraviolet penetrates and aids in plant photosynthesis and helps produce vitamin D in humans. Too much ultraviolet radiation can burn the skin, cause skin cancer and cataracts, and damage vegetation.

Unavailable Forage Forage which cannot be grazed by livestock.

Uncertainty Decision making in which there is no objective basis for assigning numerical probabilities to the different possible outcomes or in which there is no way to describe the possible outcomes. Under the Bayesian school of thought the distinction between risk and "uncertainty" is not useful and in fact may not exist. In any practical analysis all uncertain situations are assigned probability distributions, either explicitly or implicitly by the selection of the analysis method, whether or not the true probability distribution known. A probability distribution is selected for analysis purposes based on experience, intuitive feelings, knowledge, and the lack of knowledge.

Unconfined Aquifer An aquifer that has a water table; contains unconfined ground water.

Unconstrained Maximum Level of management defined as the highest possible level of a given output along with the costs associated with achieving it.

Under Plant To set out young trees or sow seed under an existing stand.

Undercut 1. In logging, the notch cut in a tree to govern the direction in which the tree is to fall and to prevent splittng.

2. In forest management, the harvesting of a quantity of timber less than the budgeted cut.

Understocking 1.Placing a number of animals on a given area that will result in under-utilizing the available forage by the end of the planned grazing period.

2. A stand of trees so widely spaced that, even with full growth potential realized, crown closure will not occur. Understocking indicates a waste of resources, for certain purposes since the site is not fully occupied.

Understory The lesser vegetation (shrubs, seedlings, saplings, small trees) within a forest stand which forms a layer between the overstory, the upper portions of adjacent trees and other woody growth, and the herbaceous plants of the forest floor.

Undesirable Species Plant species that are not desirable because they are unpalatable or low in palatability, injurious to animals, poisonous, poor stabilizers of soil and water, etc.

Uneven-Aged Management Applying a combination of actions simultaneously to maintain continuous high-forest cover, recurring regeneration of desirable species, and the orderly growth and development of trees through a range of diameter or age classes to provide a sustained yield of forest products. Specifying the number or proportion of trees of particular sizes to retain within each area, thereby maintaining a planned distribution of size classes usually regulates cutting. Cutting methods that develop and maintain uneven-aged stands are single-tree selection and group selection.

Uneven-aged Stand (All-aged Stand) A forest stand composed of trees that differ markedly in age and sizes growing together on a uniform site. By convention, a minimum range of 10 to 20 years is generally accepted, though with rotations of less than 100 years, 25 percent of the rotation period may be the minimum.

Unique Resource (Irreplaceable Asset) A resource not duplicated anywhere. A resource for which no equal (although different) resource in kind or excellence exists. Three classes of "irreplaceable" assets: One would be "gifts of individual genius" such as the work of Leonardo da Vinci or Rembrandt. A second might be called "gifts of collective genius", represented by the cultural antiquities of Egypt, Greece, and Rome, among others. Finally, we have the "gifts of nature", namely the accidents of geomorphology, biological evolution, and ecological succession. These natural endowments differ among 1. The factors of production, that are transformed through some production technology into other goods in process or into final consumption goods,.

2. And natural endowments that give rise to stimuli eliciting satisfaction in an ultimate consumption of final good sense. If a natural resource is used as a good in process, technology may provide a wide spectrum of substitutes for the particular input without significantly altering the final product either in appearance or performance.

Unit Planning The activity of developing a comprehensive general long-range plan for a planning unit.

United States Geological Survey (USGS) A bureau of the Department of the Interior USGS. was established in 1 879 following several Federally-sponsored independent natural resource surveys of the West and Midwest. The Department of the Interior has responsibility for most of our nationally owned public lands and natural resources. The USGS monitors resources such as energy minerals, water land, agriculture, and irrigation. The resulting scientific information contributes to environmental-policy decision making and public safety. For example, USGS identifies flood- and landslide-prone areas and maintains maps of the United States.

Unpalatable Species Species that are not readily eaten by animals.

Unplanned Ignition A fire started at random by either natural or human causes, or on purpose by an arsonist.

Unregulated Acres Commercial forestland not organized for sustained yield production of wood. A timber harvest may be produced in conjunction with the areas primary need. Recreation areas and their buffer zones and administrative sites are included in this category.

Unregulated This includes portions of the commercial forestland that will not be organized for regular periodic timber production under sustained yield principles.

Unregulated Land Forest land that is suitable and available, but not organized for timber production under sustained yield principles; where timber harvest is permissible but not a goal of management.

Unroaded Areas Any areas without the presence of a classified road (a road at least 50 inches wide and constructed or maintained for vehicle use). The size of the area must be sufficient and in a manageable configuration to protect the inherent values associated with the unroaded condition. Unroaded areas do not overlap with designated roadless areas.

Unsaturated Zone The zone between the land surface and the water table, containing water held by capillary forces, and containing air or gases generally under atmospheric pressure.

Unsuitable Forest Land Forest land that is not managed for timber production because: 1. The land has been withdrawn by Congress, Secretary, or Chief;.

2. The land is not producing or capable of producing crops of wood;.

3. Technology is not available to prevent irreversible damage to soils, productivity, or watershed conditions;.

4. There is no reasonable assurance that lands can be adequately restocked within 5 years after final harvest based on existing technology and knowledge;.

5. There is, at present, a lack of adequate information to responses to timber management activities; or.

6. Timber management is inconsistent with or not cost efficient in meeting the management requirements and multiple use objectives specified in the plan.

Upper Atmosphere Research Satellite (UARS) UARS is part of a long-term, international program of space research into global atmospheric change. Beginning in 1991, NASAs UARS program began to carry out the first systematic, detailed satellite study of the Earth's stratosphere, mesosphere, and lower thermosphere; establish the comprehensive data base needed for an understanding of stratospheric ozone depletion; and bring together scientists and governments around the world to assess the role of human activities in atmospheric change. Launched on September 12, 1991, UARS became the first official space component of Mission to Planet Earth.

Urban and Built-up Land and (Urban and Built-up Areas) Cities, villages, and built-up areas of more than 10 acres; industrial sites (except strip mines, borrow and gravel pits); railroad yards; cemeteries; air ports; golf courses; shooting ranges; institutional and public administration sites; and similar kinds of areas. Road and railroad acreage is included, and intermingled central acreage may be included. Farmland acreage inside city and village limits is not included. The "urban and built-up land" category takes precedence over others when the criteria for more than one category are met. Thus, residential areas that have sufficient tree cover to meet forestland criteria may be placed in the residential category.

Urban Forest All trees and associated plants and animal species that live in cities. It includes single tree streets, yards, vacant lots, and landscaped areas as well as forested areas. It may include non-tree areas like roads, trails, and ponds.

Urban Forestry 1. The practice of forestry in an urbanized environment;.

2. A specialized branch of forestry that has as its objective cultivating and managing trees in urban areas and the evaluating their contribution to the physiological, sociological, psychological (and sometimes economic) well-being of urban society.

Urban Land 1. Areas within the legal boundaries of cities and towns; suburban areas developed for residential, industrial, or recreational purposes.

2. An area so altered or obstructed by urban works or structures that identification of soils is not feasible.

3. A miscellaneous land type mapping unit.

Urban Planning (City Planning) A distinction is sometimes made between "city planning" (planning activities that take place within or at the level of a specific city boundary) and "urban planning" (planning for areas which are in predominantly urban uses but which are larger in extent than a single city's boundaries.).

Urban Sprawl The irregular and often poorly planned spreading of urban development into adjoining land areas. Many suburban residential developments are examples of urban sprawl.

Urbanized Area Such areas consists of a central city, or cities, and surrounding closely settled territory. The specific criteria for the delineation of an urbanized area are as follows: 1. A central city of 50,000 inhabitants or more; or twin cities (i.e., cities with contiguous boundaries and constituting, for general social and economic purposes, a single community) with a combined population of at least 50,000.

2. Surrounding closely settled territory, including incorporated places of 2,500 inhabitants or more; or, incorporated places with fewer than 2,500 inhabitants, provided that each has a closely settled area of 100 housing units or more; or, parcels of land normally less than one square mile in area having a population density of 1,000 inhabitants or more per square mile. An urbanized area is sometimes divided into the "central city (or cities)" and those in the remainder of the area or the "urban fringe." The central city category consists of the cities, having 250,000 inhabitants or more, named in the title of the urbanized area. The urbanized area can be characterized as the physical city as distinguished from both the legal city and the metropolitan community. Generally, urbanized areas are smaller than and in most cases are contained in standard metropolitan statistical areas (SMSA). However, in a few instances, the fact that political lines, and those of urbanized areas determine the boundaries of SMSA's by the pat tern of urban land use, means that there are small segments of urbanized areas that lie outside SMSA's. In general, then, urbanized areas represent the thickly settled core of the SMSA's, with the exceptions noted above.

Use (Allowable) The degree of use estimated to be proper until proper use is known. Forty or fifty percent of the annual forage growth (by weight) is often used as a rule-of-thumb on ranges in good or excellent condition that allows the amount of forage accelerated range rehabilitation.

Use (Capability) The first of G.A. Hills' three-level system for land evaluation. The process begins with evaluating for "use capability" based upon inherent usability potentials, then to "use suitability" on the basis of management level potentials, and finally to "use feasibility" evaluation on the basis of usability under present and forecasted socio-economic conditions.

Use (Degree Of) Utilization or consumption of weight of plant growth expressed in qualitative terms e.g., unused, slight, moderate, full, close, severe, over, extreme, destructive, etc., or as percent of weight for either an individual plant or the vegetation as a whole.

Use (District) A section of a city designated by a zoning ordinance that prescribes the use of its land and the type and characteristics of the structures that may be placed on it. The most common form of zoning is that in which "use districts" are designated (e.g., light manufacturing, commercial, single-family residential) and only the permitted use or a "higher" one is allowed. Thus in an area zoned for heavy industry, all other uses would theoretically be permitted (though there are some exceptions to this under the principle of exclusive zoning), whereas in an area of single-family residences, no other uses are likely to be permitted.

Use (Factor) An index to the grazing use that is made of forage species, based on a system of range management that will maintain the economically important forage species.

Use (Full) Maximum use that can be made of a range during a grazing season without inducing a downward trend in range condition.

Use (Nonconforming) A building or use that is inconsistent with a use district's zoning regulations. If erected after the enactment of the ordinance it may be ordered removed. If it is in being before the enactment, it may continue in use, but a new nonconforming or different nonconforming use may not be substituted. Its extension or enlargement is not permissible. Many ordinances permit rebuilding the nonconforming premises when destroyed by fire. Once the use is abandoned, however, the right to its restoration falls, and the future use of the premises must conform to the zoning. Some states allow for the abatement of such uses at the end of a prescribed and usually lengthy period of time.

Use (Over) Utilizing an excessive amount of the current year's growth which, if continued, will result in range deterioration or over grazing.

Use (Pre-existing) A land use that does not conform to a zoning ordinance but that existed before the enactment and that therefore may not be banned until its abandonment. Sometimes the "pre-existing use", if a nuisance, can be abated by court action. In some jurisdictions "pre-existing uses" have been banned after the lapse of a reasonable period equivalent to a prescribed "depreciation period" or amortization of its use by time zoning.

Use (Proper Utilization) The degree and time of use of current year's range growth which, if continued, will either maintain or improve the range condition consistent with conservation of other natural resources.

Use (Suitability) Classification based upon the degree to which a unit, in its present condition, can respond to specific management practices. In land evaluation, the process begins with evaluating "use capability" based upon inherent usability potentials, then to evaluating "use suitability" on the basis of management level potentials, and finally to evaluating "use feasibility" on the basis of usability under present and forecasted socioeconomic conditions.

Use (Utilization) Range management usage. The proportion of the current year's forage production that is consumed or destroyed by grazing animals. Range grazing, bedding, shelter, trailing, watering, etc.

Use Feasibility (Feasibility) Evaluating land based on usability under present and forecasted socio-economic conditions.

Use, Non- (Non-utilization) Lack of exercising (temporarily) of a grazing privilege on grazing lands (without loss of preference for future consideration.).

User Costs Values that users of a good or output pays to gain access or be able to use a good or service. Such costs include transportation costs, concession fees, and similar payments to other than the resource owner.

Utilitarianism A political philosophy based on the principle of "the greatest good of the greatest number". Public objective assessment of utility is possible.

Utility The usefulness or satisfaction-causing qualities of a good or service.

Utility Corridor A tract of land of varying width forming a passageway through which various commodities such as oil, gas, and electricity are transported.

Utility Theory A theory of valuation using some measure of individual preferences or intrinsic usefulness and satisfaction. The valuation can occur either in cardinal units of utility or as a comparison to other available goods given by indifference curves. An indifference curve delineates of all combinations of two goods with which an individual would be equally pleased (i.e., the area of indifferent between them). In attempting to place a value on goods that are not traded in the marketplace, proxy values can be derived by developing indifference curves between the non-market good and a good with a well-established price. In choosing among alternatives where the outcome of each alternative is uncertain, each possible outcome is weighted by its probability of occurrence. A cardinal value of the outcome of each alternative is the expected utility. The alternative with the highest expected utility is typically chosen.

UV Ultraviolet.

V

Vadose Zone The layer(s) of soil or rock between the base of a waste disposal site and the underlying water table or aquifer.

Value Added For a given enterprise, the market value of products or goods completed, less the cost of materials purchased from others.

Value Judgement A subjective opinion about the worth (e.g., good, evil, beauty) of a thing, action, or entity (even when evaluators state their lack of prejudice).

Values (Value System) The particular (frequently subjective) positive point of view of an individual or the common point of view of a group on an issue, way of life, or concerning the worth of things or experiences. Values may range from those that are subjectively meaningful to a given individual to those that are shared cultural norms. They influence the selection of the means and ends of actions, and they serve as criteria by which objects or actions are evaluated.

Van Allen belts or Van Allen Radiation belts Doughnut-shaped regions encircling Earth and containing high energy electrons and ions trapped in the Earth's magnetic field (the magnetic field has definite boundaries, and is distorted into a tear-drop shape by the solar wind). Explorer I, launched by NASA in 1958, discovered this intense radiation zone. These regions are called the inner and outer Van Allen radiation belts, named after the scientist who first observed them.

Fig. Van Allen belts or Van Allen Radiation belts

Variable 1. A quantity which may take any one of a specified set of values. It is convenient to apply the same word to denote non-measurable characteristics, e.g. "sex" is a variable in this sense since any human individual may take one of two "values," male or female. It is useful, but far from being the general practice, to distinguish between a variable as so defined and a random variable or variate.

2. Any structural element in an ecosystem that varies over time, also "component.".

3. A condition of a system in which elements or a performance measure take different values.

Variable, Driving Independent or extrinsic variables that cause the system to change, but themselves are not affected by the system.

Variable, Rate Process A variable in a simulation model which describes the physical movement of materials between state variables and which is zero when the system is at rest.

Variable, State An interactive component of a system that varies over time and the value of which at each point in time reflects the condition or "state" of that variable. When applied within systems ecology, the term refers to an ecosystem component the state of which both influences and is influenced by the conditions of other state variables in that system. Ecosystem state variables interact with each other through abiotic or biotic feedback mechanisms.

Variance (Zoning Variance) A license to act contrary to the usual rule. The granting of relief from the terms or conditions of a building or zoning law by a public agency vested with the power to authorize it. It may be given when the application of the law creates unnecessary hardship; where equally safe methods or materials may be used; where there are practical difficulties in meeting the existing requirements litreally; or where the deviation or exception would not affect substantial compliance with the regulations and not threaten health or safety.

Variety Class A particular level of visual variety or diversity of landscape character, described as Distinctive (Class A), Common (Class B), or Minimal (Class C).

Vegetation Management Management actions that change the composition or structure of plant communities including, but not limited to timber harvest, mining, livestock grazing, and fire.

Vegetation Type A plant community, assemblage, or aggregation with distinguishable characteristics.

Vegetative Manipulation Changing one vegetation type to another. It can be done by mechanical means, chemical, grazing, or fire. Usually, this is done to increase forage for livestock and can be a beneficial tool for wildlife.

Veneer Thin sheets of wood (usually less than 1/4" thick) produced by slicing or peeling a log.

Veneer Log A log of high quality and desirable species suitable for conversion to veneer. Logs must be large, straight, of minimum taper, and free from defects.

Vernal equinox The beginning of spring in the Northern Hemisphere. The time/day that the sun crosses the equatorial plane going from south to north.

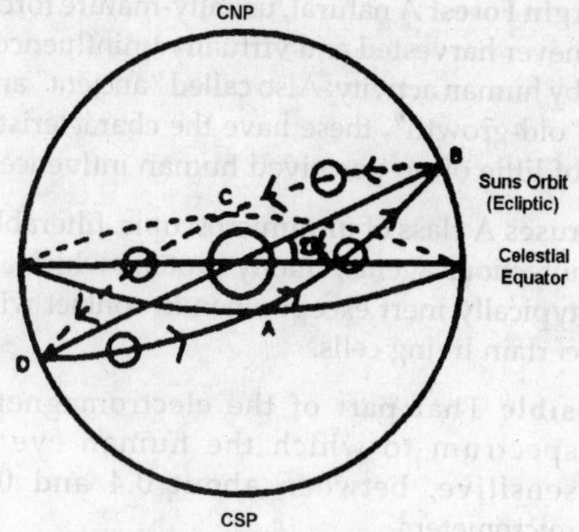

Fig. Vernal equinox

Vertebrates Animals with backbones or spinal columns. Examples include people, fish, and amphibians.

Vertical Diversity The diversity in an area that results from the complexity of the above ground structure of the vegetation. The more tiers of vegetation or the more diverse the species make-up, or both, the higher the degree of vertical diversity.

Very High Frequency (VHF) Referring to the 50-400 MHz portion of the radio frequency spectrum. Polar-orbiting satellite transmissions (APT) are made in the 136-138 MHz range using FM modulation.

Viability A population of sufficient numbers and reproductive potential to maintain its existence over time in spite of normal fluctuations in population levels.

Viable Populations A wildlife or fish population of sufficient size and reproductive potential to maintain its existence over time in spite of normal fluctuations in population levels.

Video A signal containing information on the brightness levels of different portions of an image along with information on line and frame synchronization. In the case of satellite signals, the video information is transmitted in the form of an AM modulated subcarrier.

Virgin Forest A natural, usually-mature forest never harvested and virtually uninfluenced by human activity. Also called "ancient" and "old-growth", these have the characteristic of little of no perceived human influence.

Viruses A class of ultramicroscopic, filterable, infectious agents, chiefly protein, which are typically inert except when in contact with certain living cells.

Visible That part of the electromagnetic spectrum to which the human eye is sensitive, between about 0.4 and 0.7 micrometers.

Visible Area The ground or vegetation thereon that can be directly seen from a given lookout point under favourable atmospheric conditions.

Visible/Infrared Spin Scan Radiometer (VISSR) High-resolution, multi-spectral imaging system flown on the pre-GOES-8 geostationary GOES spacecraft. Similar systems are flown on the METEOSAT and GMS spacecraft.

Visit Presence of one person upon a site (e.g., a National Wildlife Refuge.).

Visitor Day The use of an area for a total of 12 person-hours by one or more people, either continuously, or simultaneously or spread over several visits.

Visitor Opportunities The spectrum of settings, landscapes, scenery, facilities, services, access points, information, learning-based recreation, wildlife, natural features, cultural and heritage sites, etc. available for National Forest System visitors to use and enjoy.

Vista A point or area along a route of travel which affords a panoramic, unusual, or highly pleasant view.

Visual Absorption Capacity The relative difficulty or cost of achieving VQO's. It measures the land's capacity to absorb the visual impact of management activities.

Visual Character Subtype Areas of significant size and visual difference within visual character type units, e.g., steep mountain lands, foothill lands, rolling plateau lands. "Visual character subtypes" are delineated when a larger "visual character type" unit is too broad or has too great a visual character diversity to provide a logical frame of reference to classify usefully physical features.

Visual Character Type A large area of land that has common distinguishing visual characteristics of landform, rock formations, water bodies, and vegetation patterns. The "visual character type" concept and the types of land units it produces should be about the same as morphological region and physiographic regions. "Visual character type" units are used as a frame of reference for classifying the physical features of a given area as to their degree of scenic quality.

Visual Distance Zones Areas of landscapes denoted by specified distances from the observer. Used as a frame of reference, in which to discuss landscape characteristics or activities of man. The three zones are foreground (fg), middleground (mg), and background (bg).

Visual Impact The effect management activities have upon scenic or aesthetic quality.

Visual Landscape A landscape that, due to its juxtaposition with areas of human use, is vulnerable to critical observation by various segments of the public.

Visual Landscape Character (Characteristic Landscape) The overall impression created by a landscape's unique combination of visual features (such as land, vegetation, water, structures) as seen in terms of form, line, colour and texture. The naturally established landscape being viewed as visually represented by the basic vegetative patterns, landforms, rock formations and water forms that are in view.

Visual Management Program Also referred to as "Landscape Management" or "Visual Resource Management"; planning and administering the use of forest lands in such ways that the visual effects maintain or upgrade people's psychological welfare. It is the planning and design of the visual aspects of multiple-use land management.

Visual Management System A logical and orderly method developed by the Forest Service to assess the visual resource on National Forest System lands in terms of measurable objectives for specific land areas, to estimate the magnitude of the effort required to attain those objectives, and to measure accomplishment as the land management plan is implemented.

Visual Quality Objective A set of measurable goals for the management of forest visual resources. The stated goals of this visual management system are: 1. Preservation,.

2. Retention,.

3. Partial retention,.

4. Modification, and.

5. Maximum modification. Except for "preservation", each goal describes a different degree of acceptable alteration of the natural landscape based upon the importance of aesthetics. "Preservation" allows only natural ecological changes. "Retention" allows management activities which are not visually evident. "Partial retention" allows management activities which are visually subordinate to the characteristic visual landscape. "Modification" allows management activities that may visually dominate the original "characteristic visual landscape", but, when vegetation and land forms are altered, which must use the form, line, colour, texture and/or scale of that landscape for its visual characteristics. "Maximum modification" allows vegetation and landform altering management activities that dominate the "characteristic visual landscape" in the foreground and middleground but which have the same visual characteristics as the surrounding area when seen as background. Two additional short-term management goals may be required. The first, "rehabilitation", is used to upgrade landscapes containing visual impacts that do not meet the quality objectives set for that particular area. The second, "enhancement", is for landscapes having a potential for greater natural-appearing variety. Once one of the short-term goals is attained, one of the five quality goals is then applied.

Visual Resource The composite of basic terrain, geologic features, water features, vegetative patterns, and land use effects that typify a land unit and influence the visual appeal that the unit may have for visitors.

Visual Sensitivity Level (Sensitivity Level) A three-level rating system used to delineate areas receiving different amounts of exposure (present or potential) to user groups with differing attitudes towards changes in scenic quality (such as might occur as a result of management activities). The system initially classifies all travel routes, special interest areas and water bodies into areas of primary and secondary aesthetic management importance on the basis of their national importance, number of users, duration of use and area size. The system next uses the assumptions that aesthetic quality will be of major concern to recreational users and minor concern to functional users of forest areas (such as daily commuters and loggers) as the other basis for classifying the entire planning area into the three sensitivity levels.

Visual Variety Class (Variety Class) A classification system for establishing three visual landscape categories according to the relative importance of the visual features i.e., landforms, vegetation patterns, stream or lake water forms and rock formations. The class A ("distinctive") category contains those landscape features of unusual or outstanding visual quality, and, that are usually not common in the visual character type or visual character subtype. The class B ("common") category contains those landscape features which tend to be common throughout the visual character type or subtype and thus are not of outstanding visual quality. The class C ("minimal") category contains those areas with few distinguishing landscape features — including all areas not classed as A or B. This classification system is based on the premise that all landscapes have some visual values, but that those with the most variety or diversity of visual features have the greatest potential for being or attaining high scenic value.

Visual Vulnerability The degree to which human-caused changes might be seen in the landscape and their potential for degrading scenic quality in essence, the landscape's resistance (or susceptibility) to visual changes.

Volcano A naturally occurring vent or fissure at the Earth's surface through which erupt molten, solid, and gaseous materials. Volcanic eruptions inject large quantities of dust, gas, and aerosols into the atmosphere. A major component of volcanic clouds is sulphur dioxide, a strong absorber of ultraviolet radiation. Chemical interactions between sulphur dioxide and water cause sulphuric acid aerosols which can scatter some of the incident solar radiation back to space, thus causing a global cooling effect. For example, Mt. Pinatubo in the Philippines erupted in June 1991, and in the following year the global surface temperature was observed to decrease by about 0.3oC.

Volt The unit of electromotive force, or difference of potential, which will cause a current of one ampere to flow through a resistance of one ohm. Named for Italian physicist Alessandro Volta (1745-1827).

Volume Table A table of figures used to estimate the volume of wood contained in a standing tree, based on DBH and merchantable height.

Volume The amount of wood in a tree or stand according to some unit of measurement, (board feet, cubic feet, etc.) or some standand of use (pulpwood, sawtimber, etc.).

Voucher Specimen Those specimens retained in a collection as representatives of a study, and that are presumed to be accurately identified by competent authorities. Such collections are often erroneously referred to as "type collections.".

W

Warm-water Fishery Stream and lake waters that support fishes with a maximum summer water temperature tolerance of about 80 degrees Fahrenheit. Bluegills, perch and large-mouthed bass are examples.

Washery Refuse The refuse removed at preparation plants from newly mined coal.

Water Influence Zone Areas comprising existing or anticipated significant public outdoor recreational occupancy, use, and enjoyment along streams and rivers and around lakes, reservoirs, and other bodies of water, areas in which uses and activities are oriented to overwater travel and outdoor recreation. These are areas in which beauty, aesthetic values, and water quality, and usefulness are key management values, are an important part of the environment, and offer significant opportunities for existing, anticipated, or planned use and enjoyment by persons visiting or traveling through the zone. Includes areas immediately adjacent to and surrounding developed or planned sites along the banks and shores where people may fish or stroll. Comprises areas adjacent to the water along over-water travel routes and water sports use where the immediate scenic frame and foreground are subject to close scrutiny. Other foreground and backdrop areas should be recognized in other zones with appropriate management emphasis on beauty and esthetic values. Live and intermittent streams and waters where developed facilities for occupancy and use do not exist, or are not anticipated or planned, or areas which are seldom visited for recreation and enjoyment, are usually not included. Such areas should be recognized in other management zones with due consideration given to the aquatic and related wildlife habitat and to soil and water values.

Water Pollution Any substance or energy form (heat, light, noise, etc.) which alters the state of a body of water from what would naturally occur. Especially associated with those altered states which human value judgements have decreed as bad.

Water Quality The fitness of water for use, being affected by physical, chemical, and biological factors.

Water Resource Inventory (WRI) An organized characterization and evaluation of the water resource designed to address identified issues and concerns.

Water Table That surface in a ground-water body at which the water pressure is atmospheric. The highest point in a soil profile where water saturates the soil on a seasonal or permanent basis. It is defined by the levels at which water stands in wells that penetrate the water body just far enough to hold standing water. In wells that penetrate to greater depths, the water level will stand above the confining upper bed of the aquifer.

Water Table, Perched (Perched Ground Water Table) The upper surface of a local zone of soil water saturation held above the main body of groundwater by an impermeable layer or stratum, (usually clay) and separated from the main body of groundwater by an unsaturated zone.

Water vapour (moisture) Water in a gaseous form.

Water Yield 1. The volume of water runoff from a watershed, including groundwater outflow.

2. The total net amount of water produced on a wildland, including streamflow and groundwater recharge.

Waterbar A diversion ditch and or hump across a trail or road tied into the uphillside to carry water runoff into the vegetation, duff, dirch, or dispersion area so that it does not gain the volume and velocity which causes soil movement and erosion.

Waterhole An impoundment constructed or maintained to provide a water source for wildlife and other uses.

Water-related Data Any data having significance to users of water data or to those conducting water-resource projects or investigations. Examples are meteorological, water use, oceanographic, agricultural, demographic, etc.

Watershed (River Basin) 1. The total region or area above a given point on a stream or lake that contributes water to the flow at that point. It is designated by the dividing line of area from which surface streams flow in two different directions; the line separating two contiguous drainage areas.

2. The total area above a given point on a stream that contributes water to the flow at that point. The entire region drained by a waterway or which drains into a lake or reservoir.

3. British usage. The topographic dividing line from which surface streams flow in two different directions; the line separating two contiguous drainage areas.

4. The area drained by a stream and its tributaries.

Watershed Assessment An analysis and interpretation of the physical and landscape characteristics of a watershed using scientific principles to describe watershed conditions as they affect water quality and aquatic resources. Initial watershed assessments will be conducted using existing data, where available.

Watershed Condition The state of the watershed based on physical and biogeochemicalcharacteristics and processes (such as, hydrologic, geomorphic, landscape, topographic,vegetative cover, and aquatic habitat), water flow characteristics and processes (such as volume and timing), and water quality characteristics and processes (such as chemical, physical, and biological) as they affect water quality and water resources.

Watershed Integrity A watershed that maintains its characteristic diversity of biological and physical components, structure, and functional processes within its approximate range of natural variability. Watersheds with integrity display processes that manifest their characteristic structure, function, and composition. These processes include natural disturbance regimes, nutrient cycling, hydrologic functions, vegetation successlon, and species adaptation and evolution. Watersheds with integrity are resilient and capable of self-renewal within the cumulative effects of human and natural disturbances.

Watershed Management Managing all the natural resources of a watershed to protect, maintain, or improve its desired water budget, both quantity and quality, over time.

Watershed Protection A combination of complementary practices of land treatment and structural works to maintain or improve total yield, quality, stability of flow of surface and subsurface water and prevention of damage and loss due to excessive and uncontrolled runoff, flooding, salination, and siltation.

Wave 1. In electricity a periodic variation of an electric current or voltage.

2. In physics, any of the series of advancing impulses set up by a vibration, pulsation, or disturbance in air or some other medium, as in the transmission of heat light sound, etc.

Wavelength Physical distance of one period (wave repeat).

Weather Atmospheric condition at any given time or place. Compare with climate.

Weather (Meteorological Condition) The state of the atmosphere at a definite time and place with respect to air temperature, humidity, wind, precipitation, cloudiness, etc. "Climate" is long-term and generalizes trends or types.

Weather Facsimile (WEFAX) A system for transmitting visual reproductions of weather forecast maps, temperature summaries, cloud analyses, etc. via radio waves. WEFAX transmissions are relayed by NOAAs geostationary GOES spacecraft.

Weather terms.

Clear: Sky cloud-free to 30 percent covered.

Sunny: Sunshine 70-100 percent of the day.

Partly sunny and **partly cloudy:** Both terms refer to 40 to 70 percent cloud cover. Partly sunny is used in the day; partly cloudy is used at night.

Fog: A cloud on the ground. Fog is composed of billions of tiny water droplets floating in the air.

Snow: Precipitation of ice crystals.

Snow flurries: Intermittent snowfall that may result in little accumulation.

Sleet: Pellets of ice that form when rain or melting snowflakes freeze while falling. (Occurs in cold weather; hail usually occurs in summer.).

Freezing rain: Rain that turns to ice on impact with the surface.

Rain: Extended period of precipitation. Associated with large storm systems rather than single clouds or thunder storms.

Showers: Brief interval of rain that does not affect a large area.

Squall: Fast-moving thunderstorm or line of thunderstorms that often can produce damaging winds, hail, and tornadoes.

Hail: Pieces of ice that fall from thunderstorms. Hail often is composed of concentric rings of ice that form as the particle moves through "wet" and "dry" areas of the thunderstorm.

Weather warning Statement that dangerous weather is likely or is occurring. Take action.

Weather watch Statement about a particularly dangerous weather system that may occur at some specified time in the future.

Weed An unwanted plant.

Weed Cutting (Weeding) 1. Cutting made in a young forest stand, not past the sapling stage, to free the small trees from weeds, brush, chaparral, vines, sod-forming grasses, or other competing vegetation.

2. A cultural operation eliminating or suppressing undesirable vegetation, mainly herbaceous, during the seedling stage of a forest crop.

Weighting Assigning numbers to express the relative importance of items (e.g., resource yields) in a group or series under consideration usually not be directly compared by existing techniques for quantification on some measurement scale. The assigned numerical values are usually referred to as "weights" or "multipliers". Weighting is usually done according to some objective sense of the relative importance of the different types of resource yields. While weightings are usually added (i.e., additive weighting) to obtain a measure of the total social value of resource uses or resource yield mixes for planning areas, they may also be multiplied or divided (or some combination of these processes) to facilitate decision making.

Well Depth The greatest depth below land surface at which water can enter a well. For screened or perforated wells, the depth to the bottom of the screen or to the lowest perforations. For open-hole or open-end wells, the total depth.

Well-stocked The situation in which a forest stand contains trees spaced widely enough to prevent competition yet closely enough to utilize the entire site.

Wet Lines A line established by using water with or without a wetting agent, to control wildfires and prescribed fires.

Wet Meadow A meadow where the surface remains wet or moist throughout the summer, usually characterized by sedges and rushes.

Wet Sites Areas that are saturated with water and are very poorly drained to somewhat poorly drained.

Wetlands Areas that are permanently wet, or intermittently water covered, such as swamps, marshes, bogs, muskegs, potholes, swales, glades, and overflow land of river valleys. Salt marshes, and brackish marshes subject to saline and/or tidal influences are usually included. Large, open lakes are commonly excluded, but many kinds of ponds, pools, sloughs, holes, and bayous may be included. These may also be any areas that are more or less regularly wet or flooded. Where the water table stands at or above the land surface for at least part of the year. Shallow lakes and ponds, usually with emergent aquatic vegetation (such as reeds or cattails) as a conspicuous feature, are included in the definition, but the permanent waters of streams, reservoirs and deep lakes are not included. Neither are water areas that are so temporary as to have little or no effect on the development of moist-soil dependent vegetation.These are areas where the presence of water is a dominant factor and produces adverse effects on all vegetation except for aquatic plant communities.

Wetlands Lands sometimes or always covered by shallow water or which have saturated soils where plants adapted for life in wet conditions usually grow.

Wheeled Skidders An internal combustion wheeled vehicle specifically designed to skid logs from the forest to a transportation system.

Whole-Tree Removal Felling and transporting the whole tree with its crown and sometimes even its roots for chipping or trimming and cross cutting at a landing or mill.

Whorl Two to 10 or more branches growing in a ring at a node, surrounding the central leader or stem.

Wild and Scenic River Corridor The area on both sides of a wild or scenic river.

Wild River Those rivers or sections of rivers that are free of impoundments and generally inaccessible except by trail, with watersheds or shorelines essentially primitive and waters unpolluted.

Wild River Area Rivers or sections of rivers that are free of impoundments and generally inaccessible except by trail, with watersheds or shorelines essentially primitive and waters unpolluted. These represent vestiges of primitive country.

Wilderness 1. An area of undeveloped public land retaining its primeval character and influence without permanent improvements of human habitation managed and protected so as to preserve its natural condition, and which (a) generally appears to have been affected primarily by the forces of nature with the imprints of man's work substantially unnoticeable; (b) has outstanding opportunities for solitude of a primitive and unconfined type of recreation; (c) has at least 5,000 acres of land and is of sufficient size as to make practical its preservation and use in unimpaired condition; and (d) may also contain ecological, geological, or other features of scientific, educational, scenic, or historical value.

2. An area where the earth and its community of life are untrammeled by people, where people are the visitors who do not remain. In common usage, "wilderness" refers to those backcountry types of wildlands that show no obvious evidence of present or previous human uses other than foot or pack animal trails. Some use the terms primitive and wilderness interchangeably to refer to all areas displaying no obvious evidence of present or previous human uses other than foot or pack animal trails. Others make a distinction between the two, using "primitive" to refer to all areas possessing wilderness qualities and restricting the use of "wilderness" to only those portions of the larger primitive area which have officially been designated as "wilderness" areas.

Wilderness Area A congressionally designated tract of public land retaining its primeval character and influence without permanent improvements or human habitation. Management is intended to retain these characteristics.

Wilderness Condition Describes an environment suitable for becoming a Wilderness Area but one that may or may not be designated for wilderness area management.

Wilderness Study Area (WSA) One of the areas selected from an inventory of unroaded and undeveloped public lands as having apparent high quality for wilderness. They may be studied to determine whether they should be recommended for additional to the Wilderness Preservation System.

Wildfall A tree or trees thrown or their stems, broken off, or other parts (branches, foliage, fruit, etc.) blown down by the wind, or the area within which this occurs.

Wildfire (Forest Fire) Any fire other than a controlled burn or a prescribed burn, occurring on wildland. A free-burning fire unaffected by fire suppression measures. They frequently need to be suppressed, but in some areas they are allowed to burn for ecological or economic reasons.

Wildland Non-urban areas that are not intensively managed and manipulated. They include most managed forests but not city parks with their exotic plants, "manicured" lawns, and sprinkler systems. The term is not exact because it includes lands that are under management and are not truly wild. Through long usage it has come to apply to lands that are sparsely settled and present a fairly natural appearance. Forests, deserts, mountains, grasslands, and other fallow lands extensive lands are normally included. Crops, pastures, urban, residential, industrial or transportation facilities are excluded. This term is most often used as one of several terms of comparison contrasting the differing degrees of utilization and alteration of land which occur, e.g., urban lands, agriculture lands, and wildlands. The dividing lines between these states cannot be defined in any generally acceptable quantitative terms. The only firm sense of differentiation between these terms exists when they are used to describe broadly contrasting natures and intensities of land utilization. "Wildlands" are simply those natures and intensities of use on the least utilized and altered side of the continuum from totally developed to completely untouched.

Wildlife 1. Non-domesticated mammals, birds, reptiles, and amphibians living in a natural environment, including both game species and non-game species, whether considered benefi-cial or otherwise.

2. All non-domesticated and semi-domesticated mammals, birds, reptiles, and amphibians living in a natural environment, including both game species and non-game species, whether considered beneficial or otherwise.

3. Animals, or their progeny, which once were domesticated but escaped captivity and are running wild (i.e., feral animals), such as horses, burros, and hogs, are not considered to be wildlife. For some people it means undomesticated vertebrate animals, except fish. Others include wild plants, insects, and macroinvertebrates.

Wildlife and Fish Habitat System This system of the Forest Service protects and improves wildlife and fish habitat with special emphasis on threatened and endangered species. Management of wildlife and fish habitats is closely coordinated with the states because they control wildlife and fish populations. This coordination includes 1. Close working relations among National Forest, State, and private land managers;.

2. Cooperative forestry programs designed to assist land managers; and.

3. Research programs that define environmental requirements of fish and wildlife and provide management alternatives through which these requirements can be attained.

Wildlife and Fish User Day A unit used to measure visitor days, expressed as a 12-hour increment, representing use of the fish and wildlife resource for benefits.

Wildlife Habitat All of the environmental conditions of a specific place occupied by a wildlife population. The native environment of an animal, ideally providing all elements required for life and growth: food, water, cover, and space.

Wildlife Land Land and water managed primarily for fish and wildlife. Wildlife may be a secondary use of land with cropland, rangeland, woodland, etc., as the primary use.

Wildlife Plantings Agricultural crops specifically planted for wildlife in fields or small forest openings and sometimes referred to as food plots.

Wildlife Structure A site specific improvement of a wildlife or fish habitat, e.g., spring development or dugout to provide water, brushpile for cover, nestbox for birds, or rock and log placement in a stream for fish cover and pool creation.

Wildlife Travel Corridor Forested areas or other established vegetation used as travel lanes or buffer zones to connect larger stands of suitable wildlife habitat or prevent isolation of important foraging and nesting areas.

Wildlife Tree A standing dead or living tree larger than 6 inches in diameter at breast height. They may be relatively small trees and of good vigor to provide future wildlife trees in stands being regenerated today.

Wildlife, Consumptive Game and fur-bearing species harvested for sport, food, fur, study, or commerce.

Wildlife, Nonconsumptive Unharvested wildlife valuable for observation, education, enjoyment, and for its place in the wildlife community.

Willingness-to-pay An economic concept used either in non-market good valuation or when the prevailing price of a good or service is thought to be different from its true value to an individual or society. It refers to the price an individual would be willing to pay rather than be deprived of the good or service (rather than to "go without it"). Indications of willingness to pay may be obtained directly by questionnaires or indirectly by studies of payments for similar items in similar circumstances, or of related costs, such as transportation, paid to get to point of use. Validity and applicability of "willingness-to-pay" is dependent on who is assigned initial ownership rights to the good or service being evaluated.

Willy willy Australian term for tropical cyclone, hurricane.

Wind A natural motion of the air; especially a noticeable current of air moving in the atmosphere parallel to the Earths' surface. Winds are caused by unequal heating and cooling of the Earth and atmosphere due to absorbed, incoming solar radiation and infrared radiation lost to space-as modified by such effects as the Coriolis force, the condensation of water vapour; the formation of clouds, the interaction of air masses and frontal systems, friction over land and water; etc. The chart above is an abbreviated version of the Beaufort Wind Scale, named for the British admiral who invented it in 1805.

Windbreak A wind barrier of living trees and shrubs maintained for the purpose of protecting the farm home, other buildings, garden, orchard, or feedlots.

Wind chill The wind can reduce significantly the amount of heat your body retains. The following wind chill chart does not take into account such variables as type of clothing worn, amount of exposed flesh, and physical condition, all of which would alter body heat.

Windfirm Trees able to withstand strong winds, i.e., able to withstand the force of wind without being uprooted (e.g., windthrown).

Wind vane An instrument used to indicate wind direction.

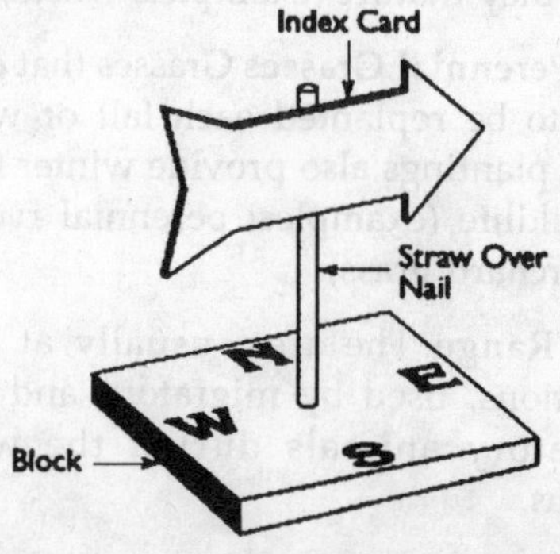

Fig. Wind vane

Wind vector Arrow representing wind velocity. The arrow points in the direction of the wind. The length of the arrow is proportional to wind speed.

Wind velocity Vector term that includes both wind speed and wind direction.

Window Term used to denote a region of the electromagnetic spectrum where the atmosphere does not absorb radiation strongly.

Windthrow (Blow Down) 1.The tendency to be uprooted by the wind. Also, a tree or trees uprooted. Trees tend to develop root systems to support themselves against the forces they face. When either logging creates a new forest edge or selected trees from a group are removed the remaining trees may face a significant increase in the force of the wind. The trees growing on the edge of a forest tend to shelter the interior trees, and trees in a group protect each other from the full force of the wind. If the remaining trees do not have root systems capable of absorbing this additional stress, they will be blown down.

2. A tree blown over.

Winter Annual Grasses Grasses that must be replanted each fall or winter. These plantings mainly provide winter forage for deer, turkeys, rabbits, and geese while they are growing but can provide seeds for birds when they mature (examples; wheat, rye).

Winter Perennial Grasses Grasses that do not need to be replanted each fall or winter. These plantings also provide winter forage for wildlife (examples: perennial ryegrass and orchard grass).

Winter Range The area, usually at lower elevations, used by migratory and semi-migratory animals during the winter months.

Withdrawn Lands Areas where action has been taken by an authority to withdraw centain lands, usually public, from uses otherwise permitted under land use or mining regulations and laws.

Wolf Tree A tree which occupies more space in the forest than its value justifies. Usually a tree which is older, larger, or more branchy than other trees in the stand.

Wooden Dollar Bib In forestry, bids above the minimum bid price but less than the sum of purchaser credits plus base rates, so-called because they do not require the purchaser to pay any more actual cash for the timber.

Woodland Land used primarily for the production of adapted wood crops and to provide tree cover for watershed protection, beautification, etc. Does not include farmstead and field windbreak plantings.

Woody Plants Plants which live longer than two years and have a thick, tough stem or trunk covered with a layer of cork.

Woody Pulp Mechanically ground or chemically digested wood (composed primarily of wood fiber) which is used in the manufacture of paper, fiberboard, etc.

Working Circle A forest area organized for a particular objective and under one set of working plan prescriptions, and embodying one silvicultural system or a designed combination of systems.

Working Group Those parts of a working circle that have the same general management prescriptions. This designation may be applied to combinations of species, types, or sites, such as pine and hardwood working groups, mixed conifer and Douglas-fir working groups, and site I-III and site IV-V working groups in some working circles. A working group is set up for areas where timber management is to be modified, such as in landscape management areas.

World Ocean Circulation Experiment (WOCE) A study of the general global circulation of the oceans. It emphasizes the measurements and understanding needed to describe and understand the circulation, to simulate it, and to predict its changes in response to climatic changes.

WWW World Weather Watch.

X

Xeric Dry conditions in a habitat rather than mesic (moderate) or hydric (wet) conditions. The nature of an organism adapted to dry conditions.

Xerosere Stages of succession in a dry area.

Xylem The tissue in higher plants which transports water, dissolved salts, and other materials (e.g., pesticides) from the roots to aerial portions of the plant.

Y

Yagi A type of receiving antenna that has several rod elements mounted on a beam. Its directional pattern of sensitivity and ease of construction make it ideal for APT direct readout stations.

Yarding Hauling timber from the stump to a collecting point and handling it there.

Yarding, Skyline (Skyline Logging) A method for moving logs to a central loading area by the use of over head cable systems that lift at least one end of the log off the ground.

Yield Table A tabular statement of goods and services (typically of a forest) expected to be produced under a specified set of conditions.

Youth Conservation Corps A human resource program which utilizes youth ages 15 to 18, to perform various agency projects. Environmental education is tied into the projects undertaken.

Z

Zephyr A Mediterranean term for any soft, gentle breeze.

Zero-Base Budgeting A management process that provides for systematic consideration of all programs and activities in conjunction with formulating budget requests and program planning.

Zone of Saturation The zone in which the functional permeable rocks are saturated with water under hydrostatic pressure. Water in the zone of saturation will flow into a well and is called ground water.

Zoning In general, separating a planning area by ordinance into zones and establishing regulations to govern the use of the land (commercial, industrial, residential, type of residential, etc.) and the location, bulk, height, shape, use, and coverage of structures within each zone.

Zoning Ordinance A local law or regulation enacted by a city council or other similar body (working under the powers delegated to it by the state) which establishes the appropriate uses within a particular, designated area (i.e., a zone) and the location, bulk, height, shape, use and coverage of structures, etc. within each zone.

Zoning Permit An official finding that a planned use or structure complies with the zoning regulations or is allowed by the granting of a special exception or variance. No new development or substantial change of use can occur until such a permit is issued in a planning area regulated by zoning ordinances.

Zoning, Aesthetic Zoning which regulates property in the interest of protecting aesthetic values.

Zoning, Agricultural Zoning which restricts land uses to agriculture and other low-density uses.

Zoning, Cluster Zoning that allows a developer to reduce his or her minimum residential lot size below the zoning ordinance's requirements if the land thereby gained is preserved as permanent open space for the community. Instead of the developer covering his or her tract with equal-sized lots, the local government alters standard lot size regulations to permit tighter patterns of house groupings. This can save money for the developer; he or she builds the same number of housing units, but does not have to clear as much land or lay as much asphalt and service facilities. In exchange, the developer leaves from 40 to 60 percent of the land open and deeds it for common use of the residents. Cluster zoning is a device for building open space into the everyday environment of residential subdivisions.

Zoning, Conservation Zoning which restrains development of unsuited areas such as steep slopes with fragile soils, areas of scenic beauty, and spaces with other natural values.

Zoning, Exclusive Use A zoning regulation that permits only one type of use in a particular established zone or district.

Zoning, Flood Plain Zoning which prohibits and restricts development within flood plains to prevent property loss, to insure public safety, and to enable the safe and natural flow of streams.

Zoning, Forest (Forest Use Zoning) Zoning which restricts land uses to forestry and related uses, e.g., timber production, watershed, Christmas tree farms, recreation.

Zoning, Large-lot Zoning which prohibits subdivision of land into individual parcels of less than a specified, relatively large lot size e.g., 1 acre or more. Large-lot zoning is used extensively to retard development.

Zoning, Natural Resource A blanket term used to refer to all zoning which restricts the uses that are made of natural resource lands so as to maintain the benefits which flow from their natural capacities, productivity or material contents. For example floodplain zoning, agricultural zoning and forest zoning are types of "natural resource zoning".

Zoning, Natural Resource A blanket term used to refer to all zoning which restricts the uses that are made of natural resource lands so as to maintain the benefits which flow from their natural capacities, productivity or material contents. For example floodplain zoning, agricultural zoning and forest zoning are types of "natural resource zoning".

Zoning, Open Space Zoning which restricts land uses so as to retain open space. It includes large-lot zoning and cluster zoning (forms of development zoning) and natural resources zoning.

Zoning, Performance (Performance Standard Zoning) Regulations providing general criteria for determining the acceptability of certain industries, land uses, and buildings as distinguished from specification standards or detailed requirements. This type of zoning does not bar an industry or use by name but admits any use, provided the particular requirements set for admission are met.

Zoning, Prohibition Any of the class of zoning regulations which only allows certain types of use in certain kinds of areas. For example, flood plain zoning frequently permits only uses which do not need structures or do not reduce the flood water storage capacity of such hazardous areas.

Zoning, Spot Designating an isolated parcel of land for a use classification that is harmful to or inconsonant with the use classification of the surrounding area so as to favour a particular owner. The courts have held such zoning to be "unreasonable" and "capricious". To avoid judicial invalidation, zoning of small areas can be enacted only when it is in furtherance of a general plan properly adopted by, and designed to serve the best interests of the community as a whole. This does not mean that an entire city or county must be zoned at one time, but there must be uniformity in the class or type of building to which the zoning applies.

Zoning, Time Restricting an existing non-conforming use within a use district to a specific period, upon the expiration of which it becomes illegal. Sometimes the time period's justification is referred to as "amortization of the use" and it is presumed to afford the nonconforming owner a reasonable number of years within which to recover his or her investment under the pre-existing use.

Zoological Area An area which has been designated by the Forest Service or others as containing authentic, significant, and interesting evidence of the national heritage as it pertains to fauna. The areas are meaningful because they embrace animals, animal groups or animal communities that are natural and important because of occurrence, habitat, location, life history, ecology, environment, rarity or other features. Areas of this type and all other special interest areas are identified and formally classified primarily because of their recreational values. Areas with similar types of values of scientific importance are formally classified as "research natural areas".

Zoology The study and classification of animals and animal life.

Zooplankton 1. Suspended or floating animal organisms usually drifting passively with water currents, e.g., protozoans, entomostracans, and various larvae.

2. The animal portion of the plankton.